AF412388

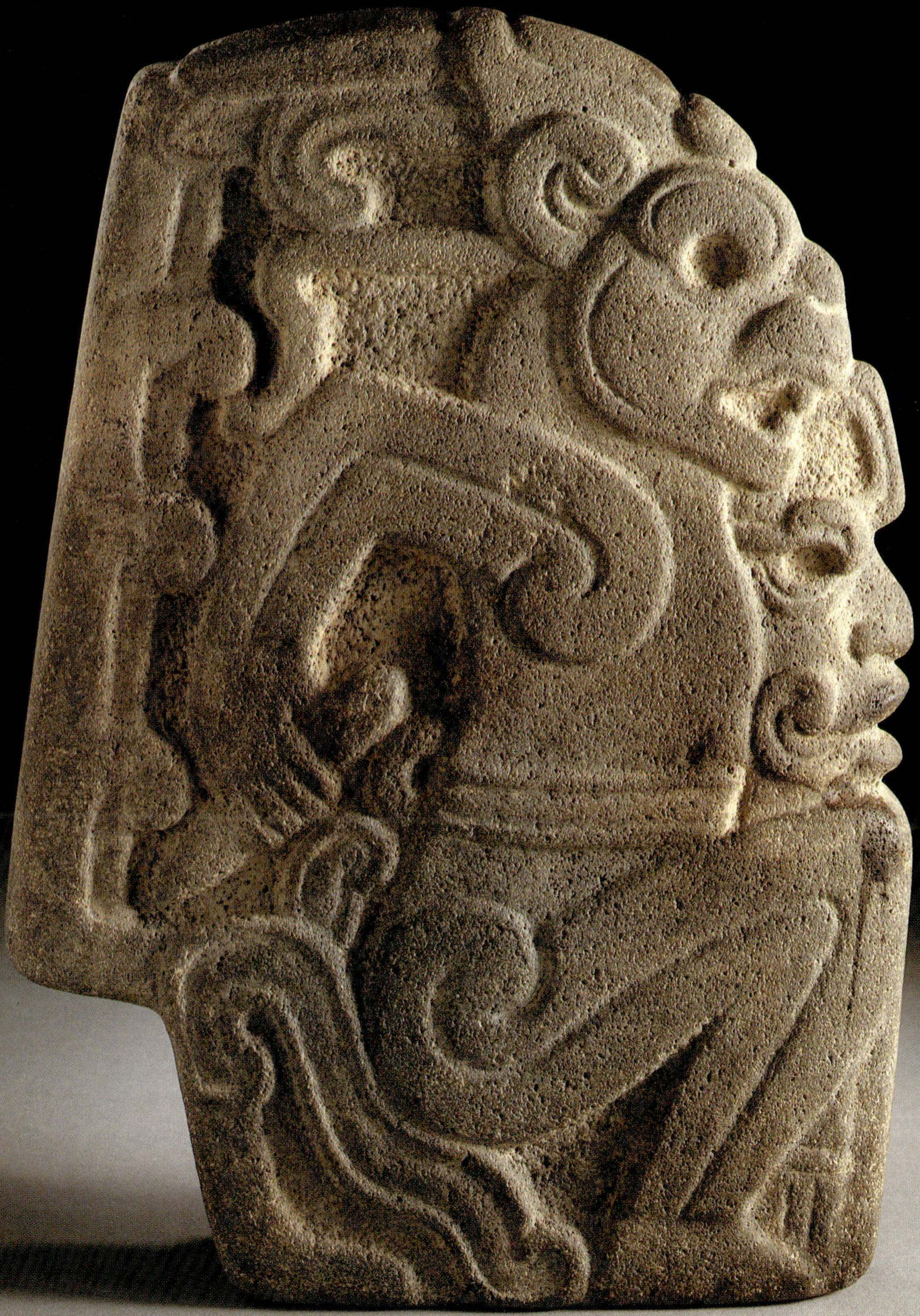

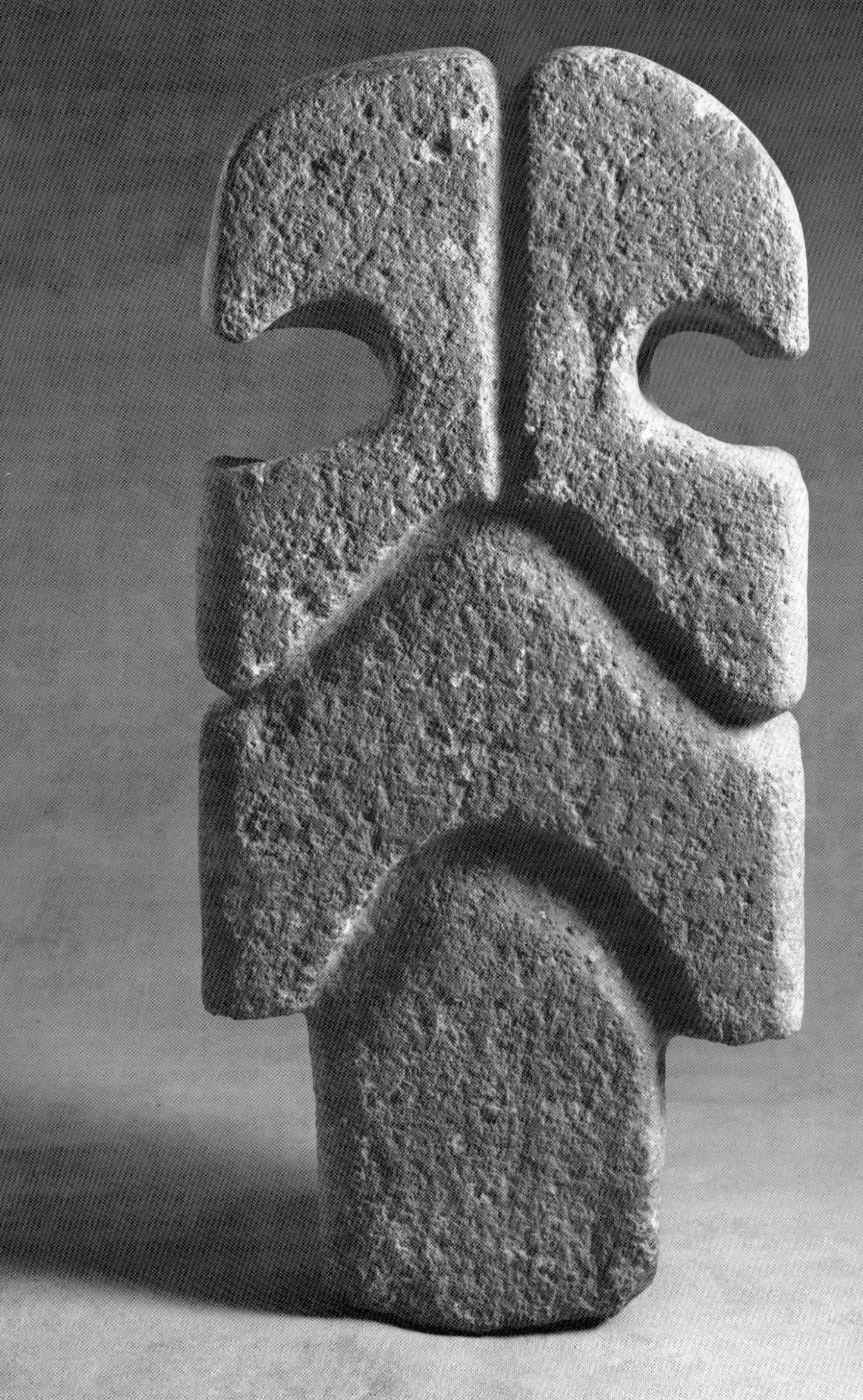

PRECOLUMBIAN FLORA AND FAUNA
Continuity of Plant and Animal Themes in Mesoamerican Art

Text by Jeanette Favrot Peterson

with an Essay by Judith Strupp Green

designed by Martha Longenecker

A MINGEI INTERNATIONAL MUSEUM EXHIBITION DOCUMENTARY PUBLICATION

made possible in part by

THE SEYMOUR E. CLONICK MEMORIAL AND SYDNEY MARTIN ROTH PUBLICATION FUND

and a grant from

THE LEGLER BENBOUGH FOUNDATION

The exhibition was funded in part by

The California Arts Council

The City of San Diego

The County of San Diego

Mingei International Corporate Associates

Mingei International Director's Circle

Library of Congress Catalog No. 90-63119
Published by Mingei International
Museum of World Folk Art

University Towne Centre, 4405 La Jolla Village Drive
San Diego, California 92122
(mailing address: P.O. Box 553, La Jolla, CA 92038)

Copyright 1990
by Mingei International Museum

All rights reserved. Printed in the U.S.A.

ISBN #0-914155-07-5

4

Cover

RABBIT HACHA
Veracruz, Gulf Coast, Mexico
Late Classic Veracruz: c. A.D. 600–900
Grey volcanic stone. H: 9¾"
The Los Angeles County Museum of Art,
Mr. and Mrs. Phil Berg M. 71.73.183
Photograph courtesy of the Los Angeles County Museum of Art

Hachas were associated with the Precolumbian rubber ballgame.

Frontispiece

HACHA/PALMA WITH JAGUAR
Classic Veracruz. Veracruz, Gulf Coast, Mexico
Late Classic: c. A.D. 600–800
Andesite. H: 13¾"
The Los Angeles County Museum of Art,
Mr. and Mrs. Phil Berg M.71.73.182
Photograph courtesy of the Los Angeles County Museum of Art

Title Page

**SERPENT SYMBOL AS
ARCHITECTURAL ELEMENT**
Huastec? Aztec? Gulf Coast or central Mexico
Late Postclassic: c. A.D. 1400–1519
Stone, andesite or limestone. H: 175⁄16"
Lent by Constance McCormick Fearing,
Photograph courtesy of the Los Angeles County Museum of Art L.83.11.908

This sculpture combines the forked tongue and rattles into a glyphic sign for
"rattlesnake." One of a pair, it was originally inserted into a pyramid or wall.

Page 9

GOPHER
Colima, West Mexico
Late Preclassic–Early Classic: 200 B.C.–A.D. 300
Burnished clay. H: 6¼"
The Land Collection ML 319

CONTENTS

DRAWINGS Joanne Heaney*

PHOTOGRAPHY Lynton Gardiner*

*except as noted

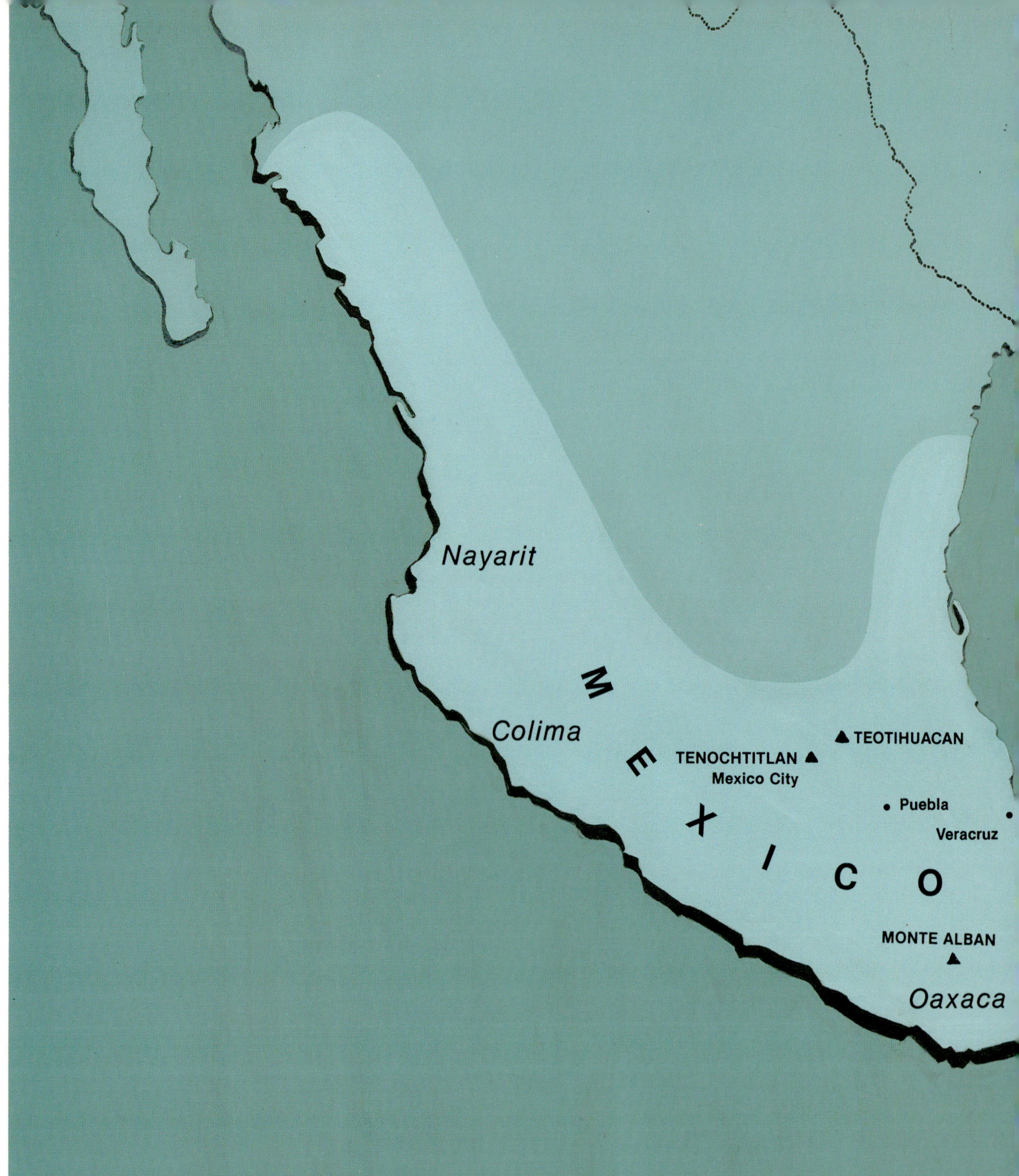

Nayarit
Colima
MEXICO
TENOCHTITLAN
Mexico City
TEOTIHUACAN
Puebla
Veracruz
MONTE ALBAN
Oaxaca
Pacific Ocean
MESOAMERICA

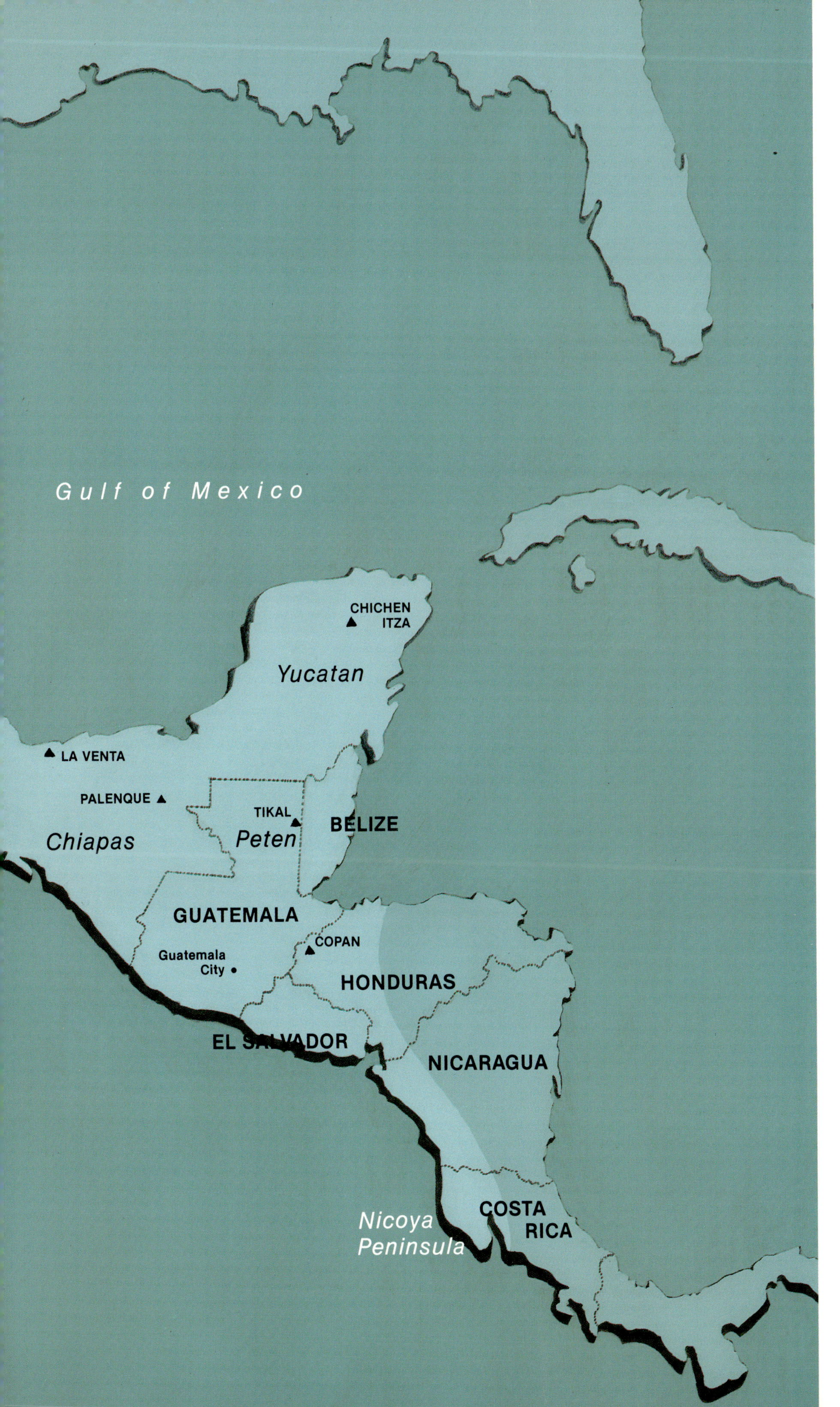

Gulf of Mexico
CHICHEN ITZA
Yucatan
LA VENTA
PALENQUE
Chiapas
TIKAL
Peten
BELIZE
GUATEMALA
COPAN
Guatemala City
HONDURAS
EL SALVADOR
NICARAGUA
COSTA RICA
Nicoya Peninsula

PREFACE AND ACKNOWLEDGMENTS

Native Americans were acute observers of the abundant flora and fauna in their environment. Fish and waterfowl were plentiful in rivers and along coastlines. Cacao (chocolate) flourished in the hot lowlands; orchids shared the same vines with the Howler and Spider monkeys in the rain forests. Deer and eagles inhabited the cooler elevations of the piney high sierra.

Many species had a multitude of uses and meanings. Certain plants and animals were transformed into symbols of religious, socio-political, and cosmological importance. Man's interaction with the natural world is expressed in Mesoamerican art from the Precolumbian period to the present day.

Parallels between plant and animal themes strengthen the concept of a culturally unified Mesoamerica, the area between Northern Mexico and Northwestern Costa Rica. Although integrated by numerous common features, Mesoamerica was not homogeneous nor was its evolution a seamless progression. Successive civilizations, from the Olmecs to the Aztecs, brought dramatic changes in the visual vocabulary and art production. The core of shared beliefs and the significance of some species remained remarkably stable. Such animals include the dog as companion of the dead soul, the jaguar as a creature of the night sun and underworld, the crocodile or saurian as the earth's surface, and the hummingbird as a metaphor for love and war.

Precolumbian fauna and flora themes lasted through the Spanish conquest to survive in the folk arts of twentieth-century Mesoamerica, particularly in terracotta and wood carvings. Traditional religious beliefs that incorporate animal lore resonate with vitality in ritual objects and dance paraphernalia.

By increasing appreciation for this extraordinarily rich art, we hope to heighten awareness of the resources this art records. As nature's gifts are abused and destroyed in Mesoamerica, many species most highly prized by Precolumbian man are at risk.

The essay on "Flora and Fauna as Metaphor in Precolumbian Mesoamerica" focuses on the diverse symbolic meanings given certain species. The flora and fauna selected are organized within the Precolumbian cosmos of upper, middle, and lower worlds. This scheme recognizes that distinctions between the spheres were not clearcut nor were the natures of the animals immutable. Fantastic hybrids, whose importance was derived from their differing capabilities, mediated between several realms. This essay introduces an immense field, complex in its implications.

Judith Strupp Green's essay on "Animal and Plant Themes in Mesoamerican Ritual and Folk Art" addresses the persistence of Precolumbian forms, materials, and motifs in Mesoamerican folk art, especially in ritual and dance. Art that illustrated plant and animal themes was often considered harmless by colonial authorities who destroyed any "pagan" work. The survival of ancient concepts is strongest in objects produced by traditional, less acculturated societies. These include the Cora and Huichol of West Mexico, the Seri, Yaqui, and Mayo of Northern Mexico, the Nahuas and Totonacs of Puebla, Veracruz, and Guerrero, the Zapotec, Huave and Mixtec of Oaxaca, and the many Maya peoples of Highland Guatemala and Chiapas.

Mingei International gratefully thanks the sponsors, lenders and all the many who have helped make possible this publication and the exhibition it documents. Museums and private individuals have generously shared their collections. All have cooperated in numerous ways and provided sustaining enthusiasm. Museum staffs contributed invaluable expertise, in particular, we would like to acknowledge the curatorial assistance of Grace Johnson in San Diego, Armand Labbe, Charles Rozaire, Nancy Blomberg, Owen Moore, and Virginia Fields of the Los Angeles area museums, Frank Norick and Joan Knudson in Berkeley, and Bob Stroessner and Teddy Dewalt in Denver. Our thanks also for the cooperation offered by Stendahl Galleries.

The director of Mingei International Museum, Martha Longenecker, provided the opportunity and vision, and the museum staff carried through their special tasks with excellence. Julia Brashares, registrar, handled her myriad responsibilities with skill and equanimity. Our gratitude is also extended to Jo Anne Heaney who made the initial exhibition design, and to the family, friends, and colleagues, especially Jan Jennings, who helped edit this manuscript.

A special thanks is owed Linda McAlister, Gordon Frost, Tom Pirazzini, and Elizabeth Cuellar for sharing with us information and photographs on folk art derived from their own research and field experience. Among colleagues in the Precolumbian field who have supported this project, I acknowledge Carolyn Tate and Jacinto Quirarte. I am particularly indebted to Elizabeth Benson, whose own substantial research in this field is innovative and who provided bibliographies and unpublished manuscripts. Heeding a caveat that Classic Maya iconography can be an area "where angels fear to tread," I sought out the scholarly commentaries of Justin Kerr, Donald Hales, and David Joralemon. My right arm throughout has been Judith Green, the principal curatorial consultant; for her insights and humor, I am most grateful.

Jeanette Favrot Peterson
Guest Curator

LENDERS

Mrs. Jarvis Barlow

Mrs. Bernice Barth

Barbara E. Busch

Mr. Frank Carroll

Gordon Frost Folk Art Collection, Benicia

Mr. and Mrs. Joseph Goldenberg

Judith Strupp Green

Sanford M. Gross

Donald Hales

The Land Collection

Linda and Steve Nelson

Dr. and Mrs. Robert Nemiroff

The Olguin Robinson Collection

Frank Papworth Estate

Tom and Alma Pirazzini

Ann E. Pitzer

Katarina Real

Dr. Norman Roberts

Martha Roth

Mr. William Schneider

Mr. and Mrs. Saul Stanoff

Stendahl Galleries

Dr. Hasso Von Winning

The Bowers Museum, Santa Ana

The Denver Art Museum

Fowler Museum of Cultural History, UCLA

The Los Angeles County Museum of Art

Lowie Museum of Anthropology, UC at Berkeley

Mingei International Museum

Museo Nacional de Arqueologia y Etnologia, Guatemala

The Natural History Museum of Los Angeles County

San Diego Museum of Man

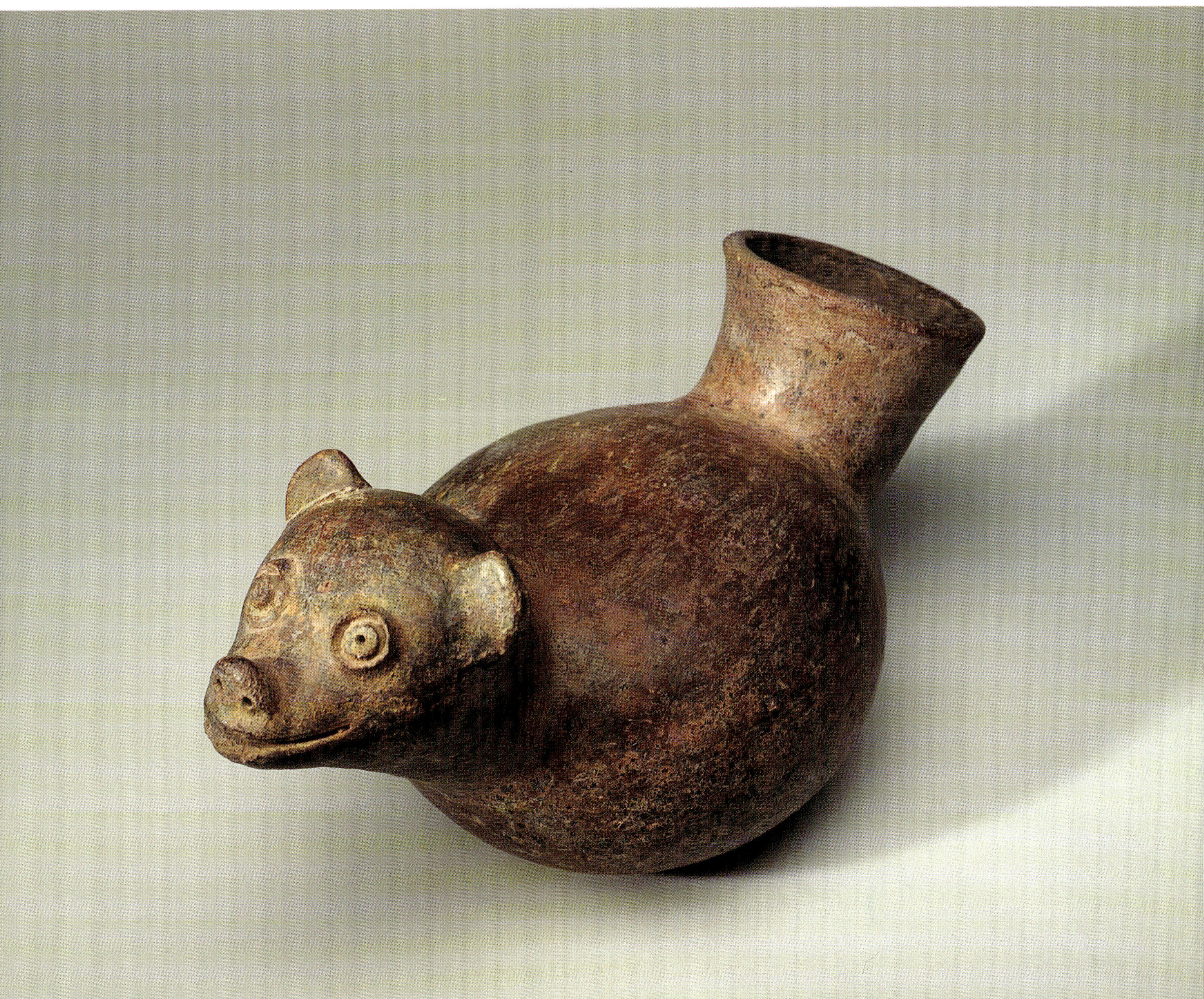

FLORA AND FAUNA AS METAPHOR IN PRECOLUMBIAN MESOAMERICA

Man has always had a sense of affinity and even identity with the animal world. Among people who interact with their natural environment on an intimate, survival level, however, this fascination is converted into a dynamic relationship often expressed in metaphorical language. Man sees reflected in nature his or her own human characteristics and behavior. A metaphor is a verbal acknowledgment for this implied likeness or analogy.

Metaphors can also be expressed visually, as is evident in Precolumbian art. The fleet-footed deer, a primary source of meat in Mesoamerica, was an animal metaphor for the hunter and warrior. The black, blood-sucking bat was symbolic for death and sacrifice. Precolumbian man used animals to understand and order natural phenomena. Conversely, nature provided models to clarify and shape the social world and cosmos. When the day sun disappeared beyond the western horizon, it traversed dark regions below the earth. As the day sun, it flew across the sky as an eagle or falcon, but as the night sun, it prowled the underworld as a fierce nocturnal jaguar.

Metaphors that draw parallels between the human and natural worlds generate ''culturally coherent bundles of meaning'' (Isbell 1985:287). In a three-way relationship, the eagle for the Aztec served as metaphor for sun and for ruler, thereby simultaneously relating the king to the solar divinity. Likewise, the image of the waterlily had many levels of meaning among the Classic Maya. The waterlily at once called up associations of the agricultural canals and the watery underworld. When the waterlily was worn as insignia by the Maya ruler, he became the provider of fertility and lord of the underworld. Among the classic Aztec, including the Nahua in central Mexico today, there exists a rich metaphorical tradition. Flora and fauna are one of several sources for metaphors which help to define not only human actions but the stages in the human life cycle. There are also metaphors for illness, curing, agriculture and earth, rulership, and founding myths.[1]

Man's dependency on the environment gave rise to a sense of awe and respect for the entire universe. In the sixteenth century, the Dominican, Diego Durán (1971:290) remarked, ''. . .even the bark of resinous trees was revered so that it would create a good fire. . .even large and small animals, fish, and tadpoles were adored and revered.'' The belief that all phenomena were animated by a life force formed a basis upon which more complex religious systems evolved.

Ancient societies structured their world view using plant and animal metaphors; they endowed their deities with powers symbolized by fauna and flora; their ruling elite borrowed from nature those characteristics they wished to emulate and control. Here we explore the selection and conversion of certain natural species into cultural signifiers by Mesoamerican man.

These ideas took form in a spectacular array of costume, architectural design, sculpture, and painting. Lost to us today are the many Precolumbian artifacts originally created in ephemeral media, such as featherwork, wood, paper, and cloth. Fortunately, we can still appreciate Precolumbian sculptures in clay and stone, including jade, their most precious material. These art works range in style from the powerful, naturalistic sculptures of the Aztecs to the esoteric imagery painted on Classic Maya vessels.

Mesoamerica encompasses most of present-day Mexico on the north and includes, moving southeast, Guatemala, Belize, and portions of Honduras, El Salvador, and northwestern Costa Rica. This vast geographic area incorporates many ecosystems. Nonetheless, Mesoamerica is considered a cohesive area culturally, bound through the sharing of such traits as the 260-day ritual calendar, the rubber ballgame, screen-style books and, most importantly, a view of the cosmos divided into three vertical layers, namely the sky, the earth, and the underworld (Gossen 1986:5–6).

Ancient Mesoamerica witnessed the growth and decline of dense populations and high civilizations. Although limited by a stone-age technology, spectacular achievements in architecture, sculpture, and painting are well known from the Olmec, Maya, and Aztec. Many other cultures from Oaxaca, the Gulf coast, and the Western region of Mexico, also produced art works of intricate craftsmanship and complexity. The period of time prior to European contact is referred to variously as Precolumbian (prior to Christopher Columbus' discovery of America in 1492), Precortesian (prior to Hernán Cortés' arrival in Mexico in 1519), Prehispanic, or simply Preconquest or precontact. The term Precolumbian has greatest currency and defines a 3,000 year time span, from about 1500 B.C. up to the Spanish conquest of Mexico in A.D. 1521. The history of Precolumbian Mesoamerica is divided into three general time periods: (1) the Preclassic (1500 B.C.–A.D. 200), (2) the Classic (A.D. 200–900), and (3) the Postclassic (A.D. 900–1521) periods. Within each of these broad time spans, further temporal subdivisions are indicated as Early, Middle, or Late phases.

CYLINDER VASE WITH WORLD TREE   1
Maya, San Agustin Acasaquastlan, Guatemala
Late Classic Maya: A.D. 700–800
Polychrome earthenware with black, red, & orange on white. H: 5½"
Museo Nacional de Arqueologia y Etnologia, Guatemala 1735
(In Clancy et al. 1985: Pl.121)
Photograph by Rolando Rosito

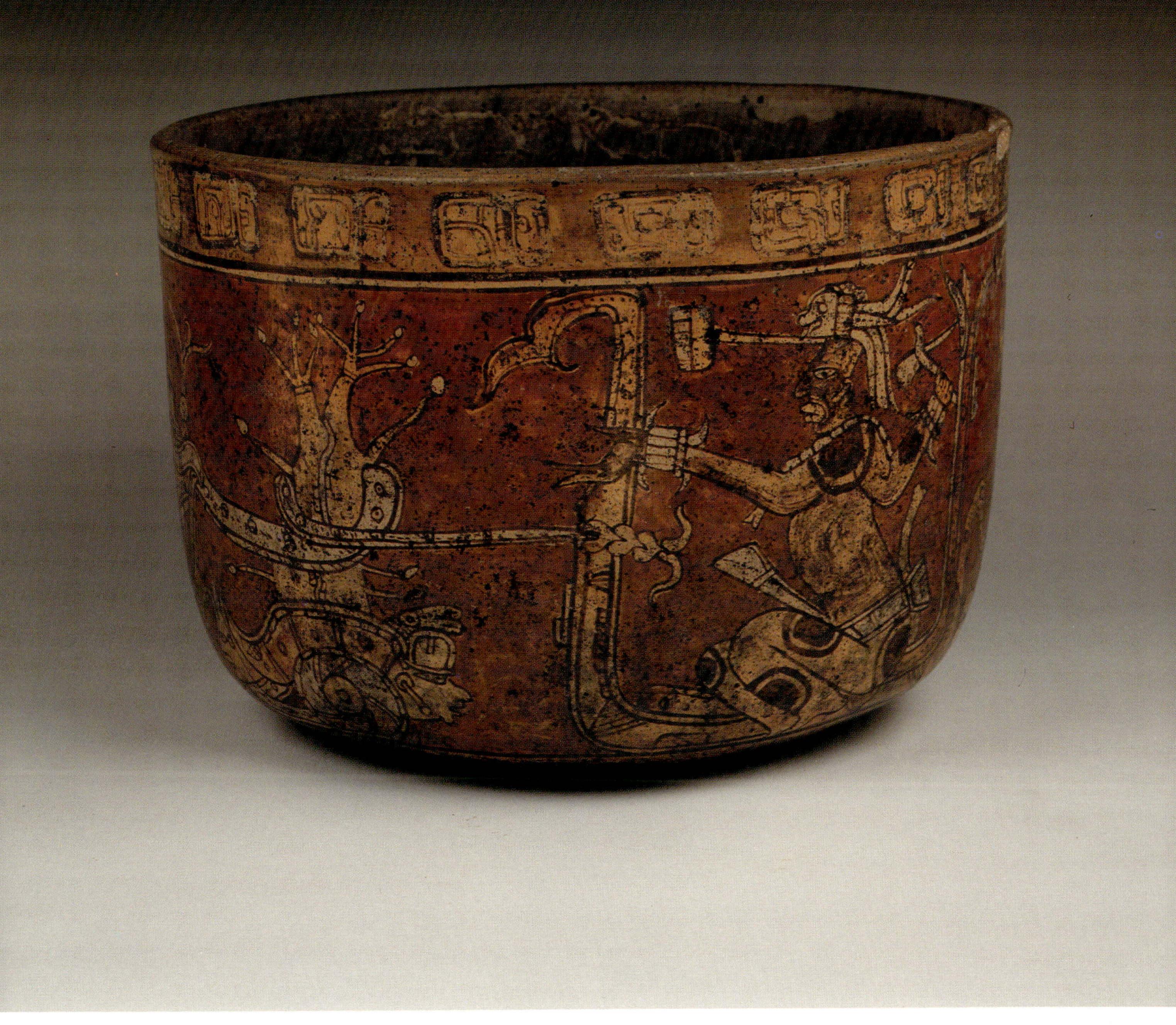

The interpretations of Precolumbian art that follow are based on a combination of archeological data, colonial documents, and ethnographic information. For this purpose, the most useful ethnographic material is collected from present-day peoples who have preserved many of their indigenous traditions. This method uses relevant material from the present as one of several aids in understanding the past. It presupposes the existence of an ancient religious system common to Mesoamerica as well as the survival of some of its components in the symbol systems of today. The arts, as visual expressions of these symbolic systems, reflect these continuities. It is also understood that many changes, some of them radical, have occurred during the rise and fall of successive civilizations. Nonetheless, enough of the ancient world view survived over the centuries to permit meaningful, if only generic, analogies with older, Precolumbian concepts. These analogies are offered as suggestions rather than definitive conclusions.

In this regard, mention should be made of the Popol Vuh, a collection of the myths and history of the Quiché Maya peoples of highland Guatemala (D. Tedlock 1985). Although colonial in date, the Popol Vuh is a partial record of a larger oral tradition that appears to have great time depth. Many scenes painted on Classic Maya ceramics are now being interpreted in light of the Popol Vuh, particularly the saga of the Hero Twins and their triumph over the Lords of death in the Underworld (Coe 1978). Relevant to this text are the Popol Vuh descriptions of animals who participated in the creation or helped the Hero Twins outwit the Underworld Lords.

THE PRECOLUMBIAN COSMOS

In Precolumbian Mesoamerica, man conceived of his earth as a horizontal plane oriented to four world directions. This earthly realm was surrounded by water and vividly depicted as a spiny crocodile or a fish-like monster floating on the sea. Although the crocodilian metaphor is no longer found in native American thinking, the fourfold division of the cosmos, with a fifth direction in the center, has persisted. The Popol Vuh records that the world was marked out with "four sides" and "four corners" (Tedlock 1985:72). Of the four world directions, the most important axis was created by the sun's daily route from east to west. The directions were not always aligned with the cardinal directions, but sometimes determined by the points at which the sun rises and sets at summer solstice (June 21) and winter solstice (December 21).

In addition to the four-sided horizontal cosmos, the universe was divided into three vertical spheres. These realms were superimposed as upper, middle, and lower worlds, sometimes expressed as the Above, the Here, and the Below. Among the lowland Classic Maya and the

Nahuas of central Mexico, the heavenly layers generally numbered thirteen, while there were nine floors or strata below the middle earthly plane (illus. a). In contemporary Mayan cosmologies, the cosmic layers are either stacked one on top of the other or they form a stepped pyramid. During any 24-hour period, the sun climbs the stair-like heavens by day to reach the summit of the sky at noon, when it begins its descent. Once past the western horizon, the sun moves down an inverted "pyramid" to reach the lowest point in the underworld at midnight.

Time and space mutually defined one another. Every day, month, year, or group of 13 years, was shaped by its location in one of the four quadrants of the cosmos. Conversely, every division in the quadripartite universe was assigned a temporal equivalent that was recorded in the ritual calendar. Since space and time were coordinated, a distinction was made in the location of the daytime and night skies. As soon as the sun set in the west, it dropped beneath the earth and entered the dark underworld, perceived as the starry sky by the Aztecs (Pasztory 1983:59). The cosmological concepts that separated day and night were also imposed on the categories of plants and animals, although, as we will see, these were not mutually exclusive realms.[2]

THE WORLD TREE

Trees, real and symbolic, acted as important cosmic metaphors. With roots below ground and a trunk supporting branches that spread skyward, a tree served as a perfect metaphor for connecting the three vertical realms. In ancient Mesoamerica, the most sacred World Tree stood at the center of the universe, or the axis mundi, often thought of as the direction up-down (Figs. 1, 2). Additionally, each of the cosmic quadrants was associated with a sacred tree, as well as a specific deity, color, and bird whose identities varied regionally. Numerous par-

---

CEREMONIAL DISH WITH   2
FOUR PARROTS ON TREE
Colima, West Mexico
Late Preclassic–Early Classic: c. 250 B.C.–A.D. 500
Clay. H: 8¾"
The Land Collection ML 227

This clay bowl with a four-branched tree and parrots has intriguing parallels with the pervasive Mesoamerican concept of a World or Cosmic Tree. It is also related to wooden pole ceremonies, including the Aztec Xocotl pole, described as supporting a green bird on four branches (Durán 1971:204). (In Dwyer & Dwyer 1975: fig.49; Nicholson & Cordy-Collins 1979: fig. 77a).

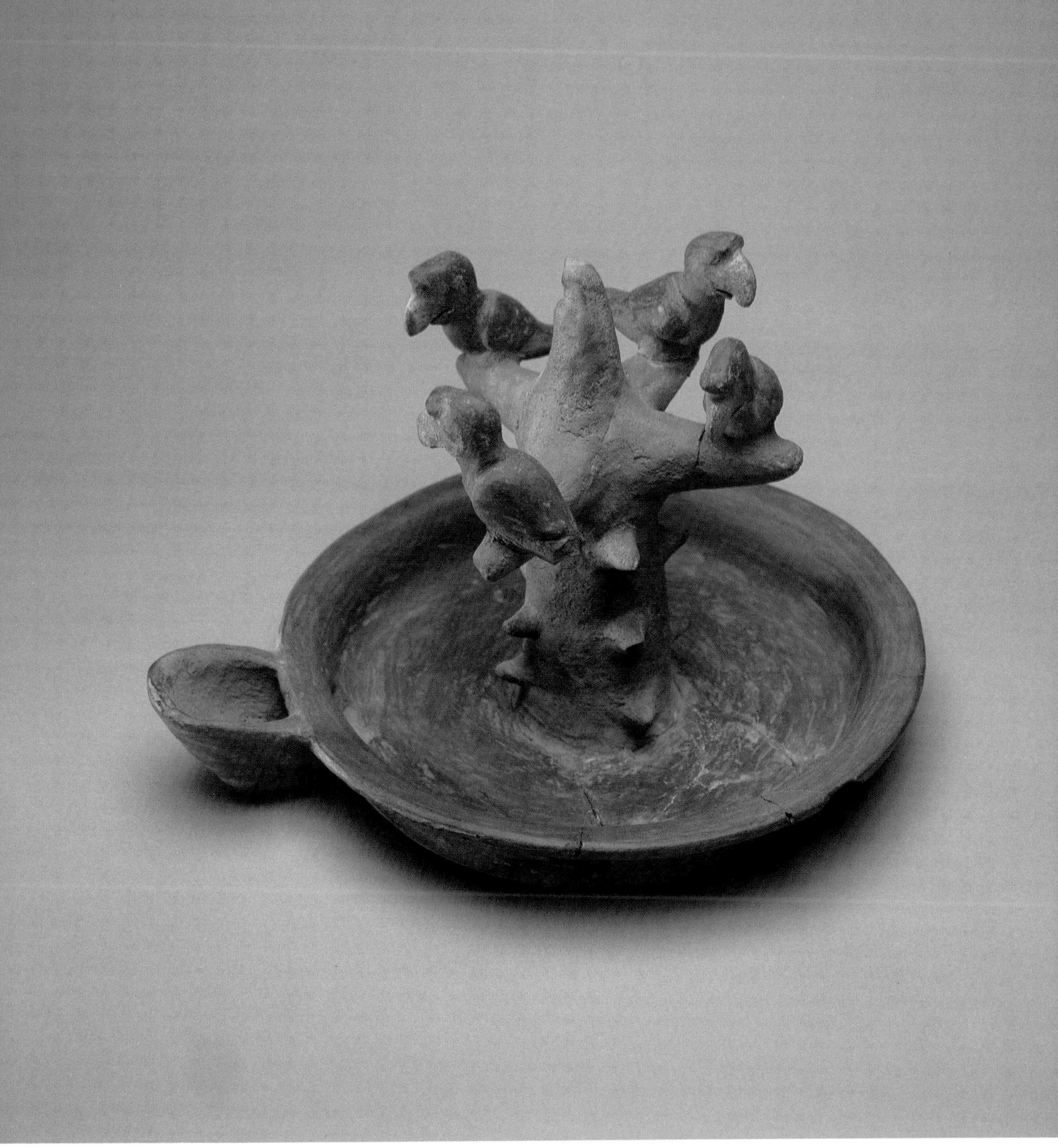

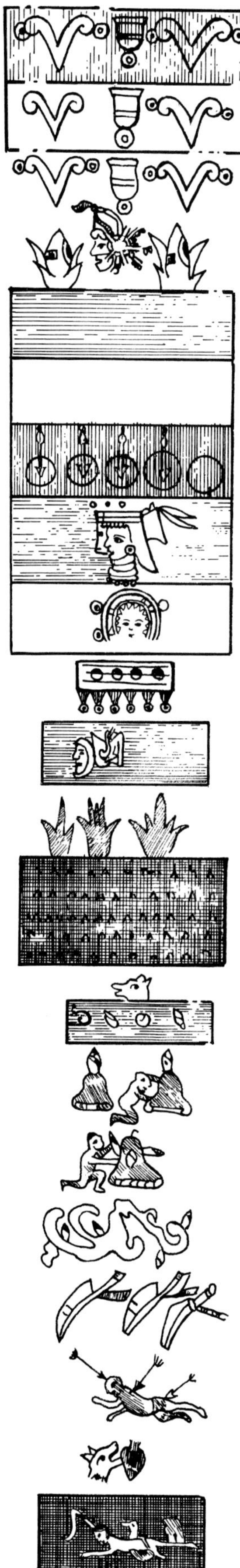

allels exist between Precolumbian imagery and contemporary native cosmologies. There are shared beliefs in World Trees as metaphors for ancestry, abundance, and rulership. These similarities are reinforced by rituals involving wooden poles that survive to the present day.

Sacred trees were located at the four corners and center of the world as depicted in Postclassic pictorial manuscripts. The clearest example of directional trees is found in the Maltese cross cosmogram in the *Codex Fejérvary-Mayer* (illus. b). Each of the four trees grows out of a terrestrial symbol and is surmounted by a supernatural bird. The world tree in cross form was an image that had precedents in Mayan relief sculpture. At the Classic Maya site of Palenque, an eighth-century Maya ruler, Pacal, was buried in a stone sarcophagus or tomb. The sculptural design on the lid of this sarcophagus depicts a cross-like tree surmounted by a celestial bird (Schele & Miller 1986: Pl. 111). The tree acted as a passage between the land of the living and the dead. Through this "Tree of Life," Pacal's soul could ascend to the eternal life of a deified ruler. A Maya king maintained order in the cosmos; this pivotal role was declared through World Tree symbols that appeared on his scepter, apron, and headdress.

In Mayan creation myths, the "first tree" played a mythical role as a primordial source of life. In addition, cosmic trees delineated and supported the heavens. Today, trees help define the supernatural and social space of some Mayan villages. Trees (or stone columns) mark the center and four entries to the village. Directional trees are also called trees of abundance, linking concepts of creation and fertility. Numerous Maya peoples trace their lineages to specific trees. Trees not only begat founding ancestors but permit dead souls to move from one world to another. It is not surprising that Spanish friars, like the Franciscans in Yucatan, moved quickly to channel the powers of the ancient cross-like tree into the symbol of the Christian cross.

As natural metaphors of the World Tree and ancestral "family trees," two important trees were singled out for

---

THE AZTEC VERTICAL COSMOS   a
Thirteen sky levels and nine layers of Earth and Underworld. *Codex Vaticanus A,* fols. 1v. & 2r. (Adapted from Nicholson 1971:407.)

XOCOTL POLE IN AZTEC CEREMONY   b
Detail from tenth trecena in *Codex Borbonicus.*

THE AZTEC HORIZONTAL COSMOS   c
Maltese cosmogram with four directional trees and four birds. *Codex Fejervary-Mayer* (Page 1). (Modified design from Leon Portilla 1963:47.)

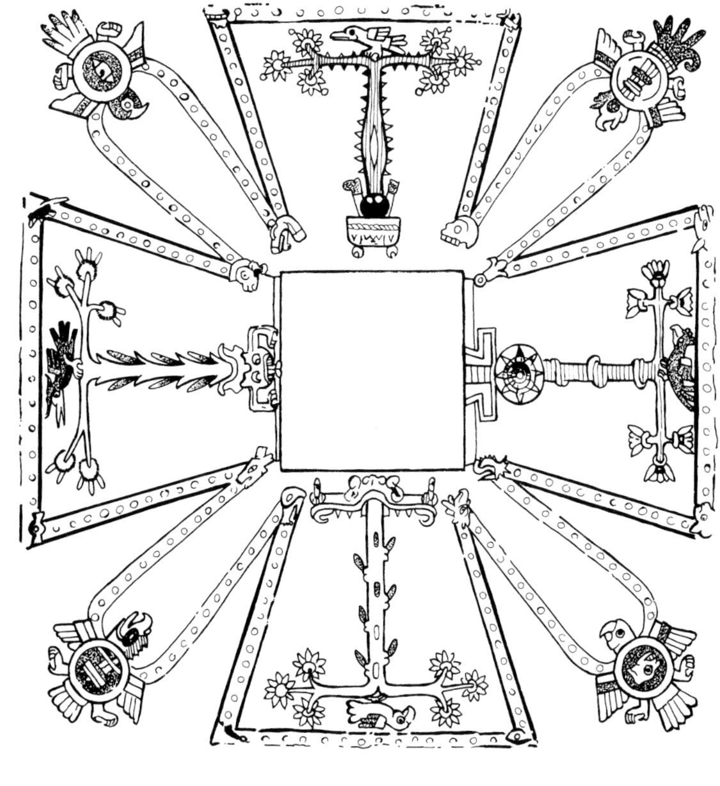

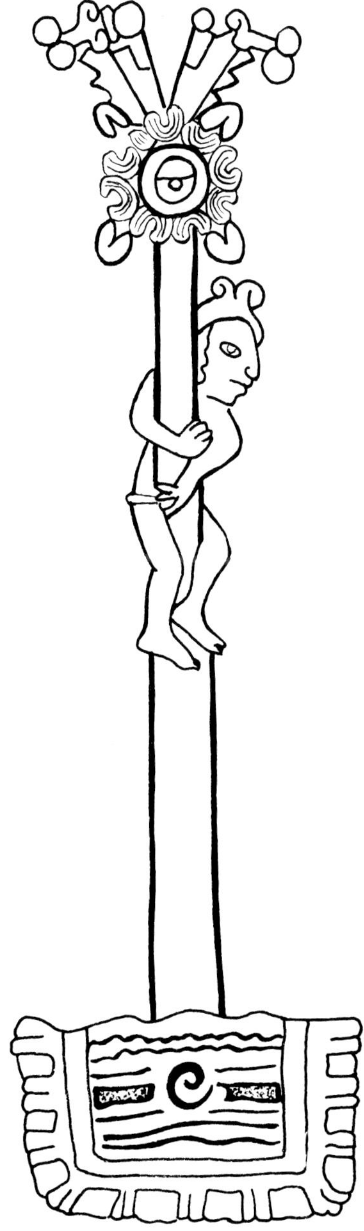

their size and longevity. These species were the *ceiba* among the Maya and the *ahuehuete* among the central Mexican peoples. The ceiba or giant silk cotton tree (Seiba pentandra), is called the *yaxche* (green tree) or *yax imixche* in Yucatec Maya. The giant ceiba can attain a height of 140 feet and a trunk diameter of 8 feet. The red color of the ceiba's new leaves (Seiba pentandra) and sap (Seiba pochote) are appropriate to its associations with the blood lines of a community (Schele and Miller 1986:277–278). The tree can also seemingly "rejuvenate," as branches of the ceiba take root quickly. Like the ceiba in its dimensions and age, is the *ahuehuete* or cypress (Sabina). Both trees offer generous shade in the umbrella-like extent of their branches; this shade was compared to the protection offered by a great ruler.[3]

The World Tree is an integral part of ceremonies that feature wooden poles, such as pole-climbing rituals and the *Juego de volador* or Flying Pole ceremony, that will be elaborated in the next essay. We can assume great time depth for the widespread pole-climbing rituals. When performed, they are still related to the concepts of a central cosmic axis, ancestral lineage, and fertility.

Precolumbian precedents for aspects of this ritual are found in the Aztec ceremony of Xocotl Huetzi (Fruit falls down). A large shorn pine tree was raised in the temple courtyard and surmounted with a corn dough image. After a ritual of human sacrifice, food offerings, and dancing, sons of lords and chieftains scrambled up the pole to reach the corn dough idol and cast it down to the crowd below. Multiple meanings of the Xocotl Huetzi ceremony include that of harvest plenty. Since the pole climbers were honored nobility, this ceremony also validated the divine right to rule among the Aztec elite. The cosmic significance of the Xocotl pole, bridging the up-down space between the earth below and the sky above, is suggested in the *Codex Borbonicus,* a sixteenth-century native-style pictorial manuscript from central Mexico (illus. c). In the *Codex Borbonicus* image, the Xocotl pole is planted in a watery trough and is capped with a "god's eye" symbol for the night sky. Precolumbian tree-pole ceremonies may also be represented in ancient West Mexican clay sculptures of figures on poles from Nayarit and a four-branched tree with parrots from Colima (Fig. 2).[4]

In this essay, the more important of the myriad flora and fauna in the Precolumbian menagerie are arranged according to the cosmic divisions of Sky, Earth, and Underworld. Between these larger cosmic sections are transitional zones for the species that most commonly mediated between the two adjacent realms, i.e., Sky to Earth (serpent) or Earth to Underworld (dog; water bird). Although the use of a cosmic scheme is an organizing format, it also provides one means for understanding the Precolumbian bestiary (cf. also Furst 1978:27–28). Certain plants and animals were selected as visual equivalents of this universe — as metaphors for celestial bodies, the surface of the earth, or the underworld realm.

Precolumbian man did not compartmentalize his dynamic universe. Distinctions between realms were fluid. Certain animals were associated with two levels at once; others were unique for their seeming ability to transgress and transcend boundaries. Such species as the cormorant flew with the celestial bodies and also dove deeply into the underworld realm of water. Others, such as reptiles, dwelled in two cosmic zones by swimming underwater and moving onto land with equal dexterity.

The paradoxes and ambiguities in nature were also accounted for by pairing and even fusing contradictory characteristics. Accordingly, hybridized mythical beasts that combined multiple animal traits were very prominent. The Olmec Dragon, one of the earliest of the fantastic supernaturals, had the jaws of a crocodile, a jaguar's fangs and nose, serpent body, bird's wing, and some human features. Such composite animals were all the more powerful for simultaneously representing the features of several cosmic realms.

The headings given each type of plant or animal in this essay are in English and Spanish. Italicized are their most common names in Nahuatl, the Aztec language, as well as in Yucatec Maya, one of more than 35 Maya language groups.

POLYCHROME URN WITH EAGLES   3a, 3b
(Two views of same urn.)
Aztec-style. Veracruz (?), Mexico.
Late Postclassic: c. A.D. 1400–1519
Polychrome clay. H: 41½"
UCLA Fowler Museum of Cultural History, 75.01531
Photographs courtesy of FMCH

In Postclassic Mexico, the three-way metaphor linking eagles with the sun and the warrior class underscored the imperialistic ambitions of the Aztec empire. On this voluminous urn, eagles alternate with disks symbolizing the sun. The striding eagles hold sacrificial banners and the solar disks contain the symbol for ollin or movement, denoting the current epoch of Aztec history. This type of clay urn was made in Cholula, Puebla, and was widely traded.

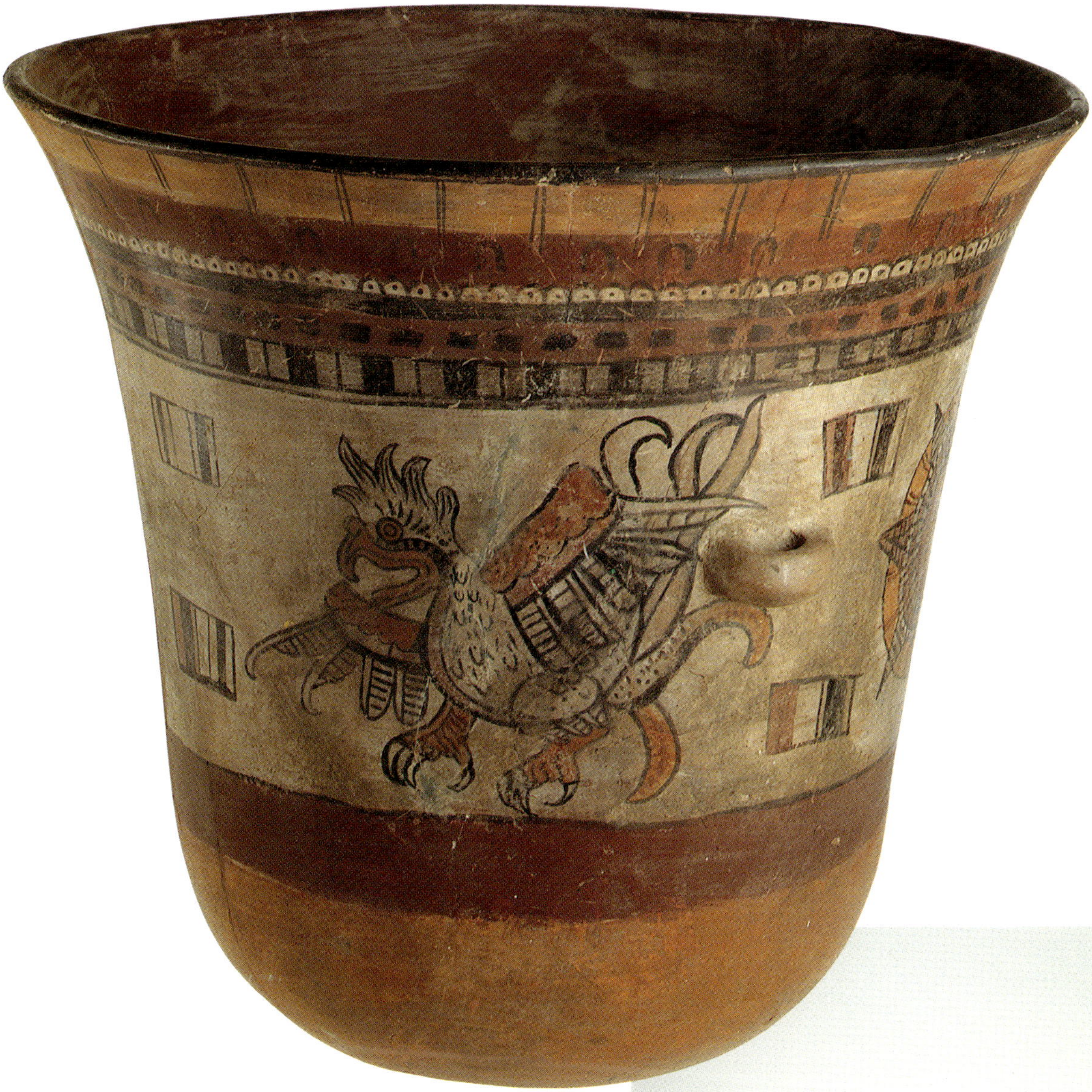

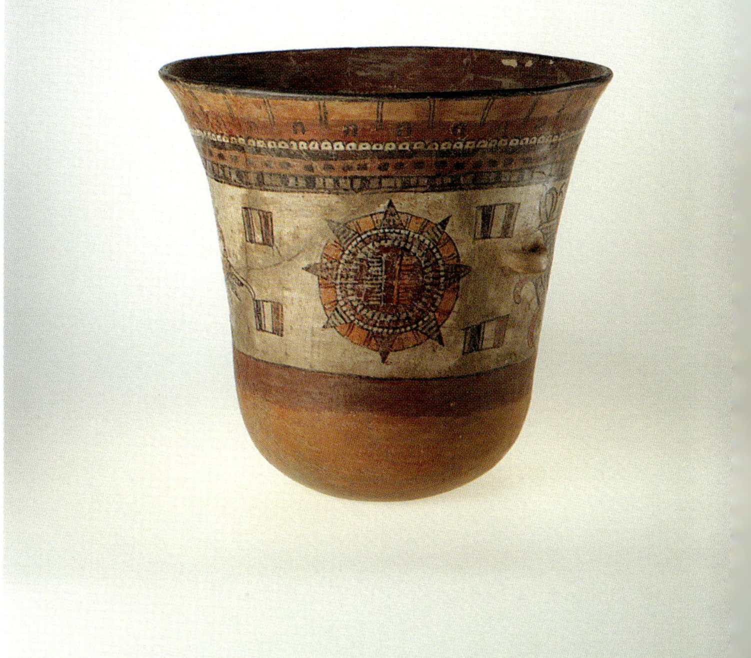

SKY

Sky animals were associated with the celestial sun, moon, and Venus, where they acted as messengers between man and his deities. Logically, these included many of the birds, with particular importance assumed by the eagle, parrot, and hummingbird. Each of the thirteen celestial deities had winged disguises, all were birds with the exception of the butterfly. Each bird corresponded to one of the thirteen heavenly strata and included the quetzal, turkey, quail, and hummingbird.

In the Precolumbian cosmos, the day sky was overhead and the night sky in the underworld. This distinction did not seem to apply to the inconstant moon, whose waning and growth took place in either realm. The day sun was located in the sky but traveled below the earth as the night sun. Thus, the birds of prey with daytime hunting habits, such as the eagle, hawk, and falcon, were solar creatures of the overhead day sky. Night hunters, such as owls, bats, and jaguars, were almost always placed within the subterranean realm of the underworld. While birds made up the largest number of sky animals, reptiles like the two-headed serpent, were also attributed solar associations. In spite of their strong ties to the earth's fertility, deer and rabbit are included in the sky section because of their symbolic ties with the sun and moon respectively.

BIRDS OF PREY

The soaring flight patterns of birds of prey made them excellent metaphors for the sun. Their speed, visual acuity, and powerful talons, were also models for the successful hunter and warrior. The hawk (Accipitridae) family, which includes the eagle and vulture, and the falcon (Falconidae) family share so many features that they were often interchangeable in Prehispanic art and myth. These birds are characterized by sharp curved claws and hooked beaks. Most can fly to impressive heights, then make dramatic dives to pursue their prey.

These were the animals thought best suited to communicate with the celestial gods. Falcons in the *Popol Vuh* were messengers for celestial gods because of the ease with which they could fly between the three realms. "And for this falcon it wasn't far to the earth here, nor was it far to Xibalba; he could get back to the sky, to Hurricane, in an instant" (Tedlock 1985:105). The heights to which eagles and hawks can fly is indicative of their special mystical powers to "see and hear everything" (Lumholtz 1902, II:7–8). Such powers were not only adopted by the shamans, as they are among the Huichol peoples of Western Mexico today, but they were often appropriated by the Prehispanic ruling class.

EAGLE — AGUILA   *cuauhtli; kunkuk*

The eagle served as one of the Aztecs' most fundamental emblems for the sun's disk and path. In their mythology, eagle was said to have been singed black by the sun, and the sun itself was referred to as an "ascending eagle" or "soaring eagle." As the supreme solar deity, the eagle was featured in the founding of the Aztec capital of Tenochtitlan. The eagle on a nopal cactus with a serpent in its mouth was an imperial logo that symbolized Aztec hegemony over sky and earth. It was adopted as the Mexican national emblem and is the central image on the flag of Mexico.

The eagle was also an emblem for one of the elite warrior organizations. Paired with the jaguar warriors, these two military groups invoked the dominant creatures of the sky and underworld. Eagle warriors were considered "servents of the sun." Aztec terms for fearlessness and warfare were given eagle qualifiers. Since these qualities were incarnate in the Aztec ruler, he was likewise compared to an eagle. The eagle-solar-warrior metaphor became very closely affiliated in Postclassic Mexico (Figs. 3a, 3b).

In their role as "peoples of the Sun," the Aztecs felt blood sacrifice was necessary to sustain not only the sun but to keep the entire universe in motion. Eagle warriors were expected to give their lives as part of this mission. If killed on the battlefield, eagle warriors were promised an afterlife in the celestial heaven of the sun. Eagles were sometimes represented holding sacrificial human hearts in their claws. The Aztec stone receptacle for these hearts was called *cuauhxicalli* or "eagle vessel." Stylized eagle feathers that were carved on such vessels served as shorthand symbols for the sun.

HARPY EAGLE

The role of the eagle in blood sacrifice appears early in Mesoamerican history. The formidable Harpy eagle (*Harpia harpyja*), a lowland forest dweller and the largest eagle in the Americas, was a potent symbol for the Preclassic Olmec and Classic Maya. Airborne, these

---

HARPY EAGLE BLOODLETTER   4
Olmec. Gulf Coast, Mexico.
Mid Preclassic: c. 900–500 B.C.
Black jade. H: 6⅛"
Private collection

The Harpy eagle decorating this Olmec ritual tool is identified by its crest, prominent knob above a hooked beak, and large claws. The bloodletter was used for piercing body parts for sacrificial blood which was thought necessary to promote agriculture and sustain life. The incised design on the implement's back is related to its sacrificial function (illus. d). A three-lobed motif that resembles sprouting maize is above a highly geometric four dots and bar symbol for the Olmec earth deity.

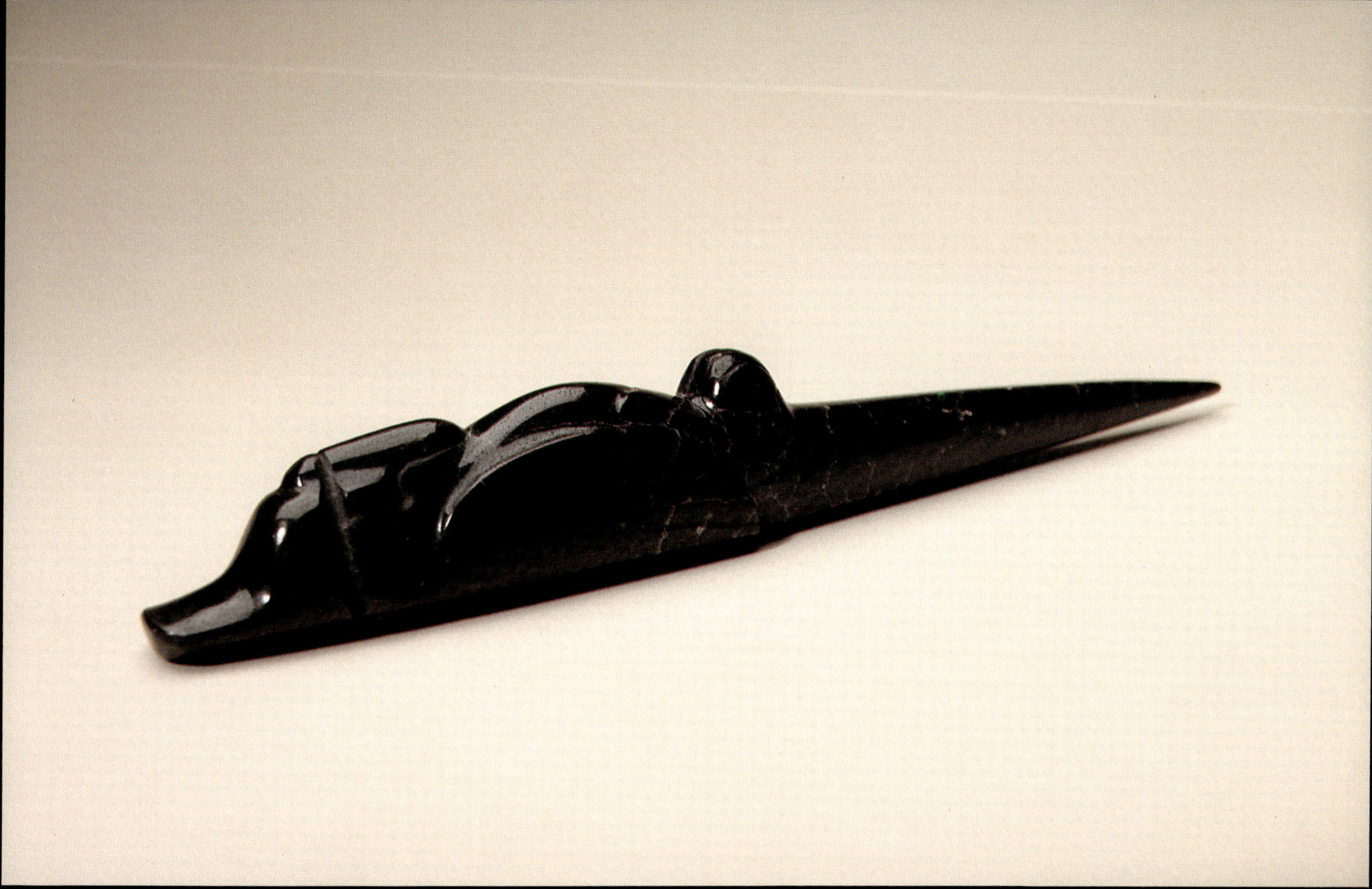

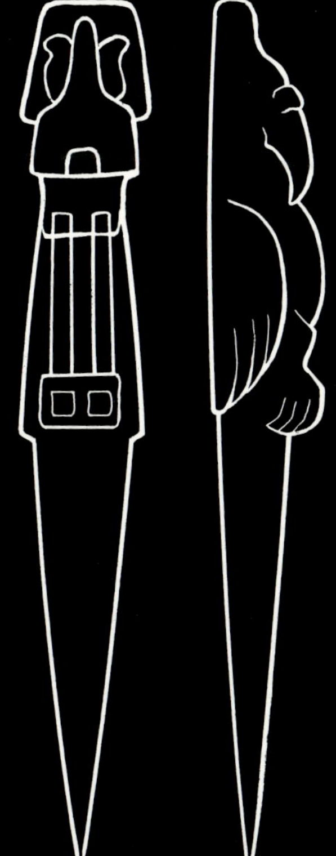

eagles can fly beyond human sight. Their platform nests are built as high as 150 feet off the ground. The Harpy eagle can maneuver with great agility through forest branches to catch even monkeys and opossums. Harpy eagle traits are incorporated into several Olmec super-naturals, most prominently in the Bird Monster (god III) as defined by Joralemon (1976). The Bird Monster manifests itself in a range of images, from a full figure eagle effigy to a complex hybrid creature that also includes reptilian features.

A Harpy eagle decorates the upper portion of an Olmec black jade bloodletter from the Gulf Coast region (Fig. 4). Stone perforators were one of several types of blood-letters, including natural dagger-like implements such as cactus thorns, sting-rays spines, and sharks' teeth. They were used for self sacrifice, practiced by piercing the ear, tongue, or penis. Ritual blood sacrifice was a form of propitiating the gods and insuring their continuing gifts to man. The relationship between blood sacrifice and the earth's abundance is referred to in an abstract symbol for corn and the Olmec earth deity incised on the back side of the Harpy eagle bloodletter (illus. 4). The images and function of this jade bloodletter convey the same message of fertility.

VULTURE — ZOPILOTE   *tzopilotl; chom*
KING VULTURE – BUITRE REAL *cozcacuauhtli; kuch*

New World vultures as a group share celestial and solar associations. Among the Mopan and Kekchi peoples of the Maya region, the King vulture in particular is con-sidered the sun's rival for the moon; it is the vulture who carries the moon away on his back. The all-black Turkey or Black vulture was thought to be able to obscure the sun with its wings. According to Maya legends, this vulture was originally endowed with spectacular blue and yellow feathering. However, the gods punished the vulture's vain doting on his plumage by sending it into the sun's rays. There it was singed coal black. The bird was also thereafter required to eat carrion or dead animals.

The King vulture (Sarcoramphus papa) was generally represented with a distinct fleshy knob on its upper beak (Fig. 5). This fleshy protuberance can take the form of a three-petalled flower, which is how the Classic period artists of Oaxaca portrayed it in the headdress of a Zapotec bird deity. Like another dark, winged creature, the bat, all vultures were also connected with human sacrifice. It is likely this sacrificial aspect arose from their consumption of both living and dead flesh. In the *Dresden Codex* (p. 3), the vulture's sacrificial role emerges. The bird sits on top of a tree growing from the abdomen of a sacrificial figure; in its beak, the vulture holds the victim's eyeball (Seler 1909:51). They are still said by the Maya to gorge out men's eyes, although the ancient rites of sacrifice have become acts of vengence.[5]

HUMMINGBIRD — COLIBRI   *huitzitzilin; tsunuun*

Delicate, irridescent hummingbirds were also associated with the sun and warfare. In fact, hawks and eagles were, and are, often paired with hummingbirds in Meso-american mythology. Many of the over 300 species that make up the New World hummingbird family (Trochilidae) cluster around the equatorial region. The hummingbird's spectacular shiny feathering and aerial acrobatics make it a perfect solar metaphor. Since hummingbirds are capable of hovering motionless and fly up, down, and even backwards, these flight patterns are related to the sun's ''hovering'' in the sky at the solstices when it also reverses its celestial direction (Hunt 1977:68).

Most unique is the hummingbird's apparent ability to die and resurrect. When cold, during the night or for short periods during the winter, a hummingbird can assume a torpid state. For the Aztecs, it appeared to come to life again in the spring at the very time of renewed life and planting. These phenomena were also analogous to the sun's ''rebirth'' or rising every morning in the east. Powers of life and fertility were transferred to the Aztec deity of war and sacrifice, Huitzilopochtli (Hummingbird-Left).

The hummingbird's long pointed beak is also a carrier of sexual connotations as it is plunged deeply into the corolla of flowers to suck the nectar. Among numerous Mayan beliefs concerning the sun as the moon's lover, the sun either wears a hummingbird disguise or turns into a hummingbird. In some Maya myths, the young night sun as a hummingbird sucks the honey from the five-petaled flower or moon. In others, the hummingbird-sun is killed and subsequently revived by the moon herself.

Although seemingly vulnerable, hummingbirds are known to be territorial and aggressive fighters. Their sharp beaks were not only affiliated with weapons but with sacrificial implements. Olmec bloodletters were created in hummingbird form, although jade humming-birds were also used as insignia or costume elements

---

ANTHROPOMORPHIC   5
VULTURE WHISTLE
Maya. Jaina, Campeche
Late Classic: A.D. 500–800
Clay. H: 5¼"
Collection of The Denver Art Museum, gift of Cedric Marks 1971.402

This clay figure, possibly a priest, wears a vulture mask with the characteristic wrinkled forehead and knob above its long curved beak. Vultures were primarily solar sym-bols and, because of their habit of eating dead meat, were also associated with ritual sacrifice.

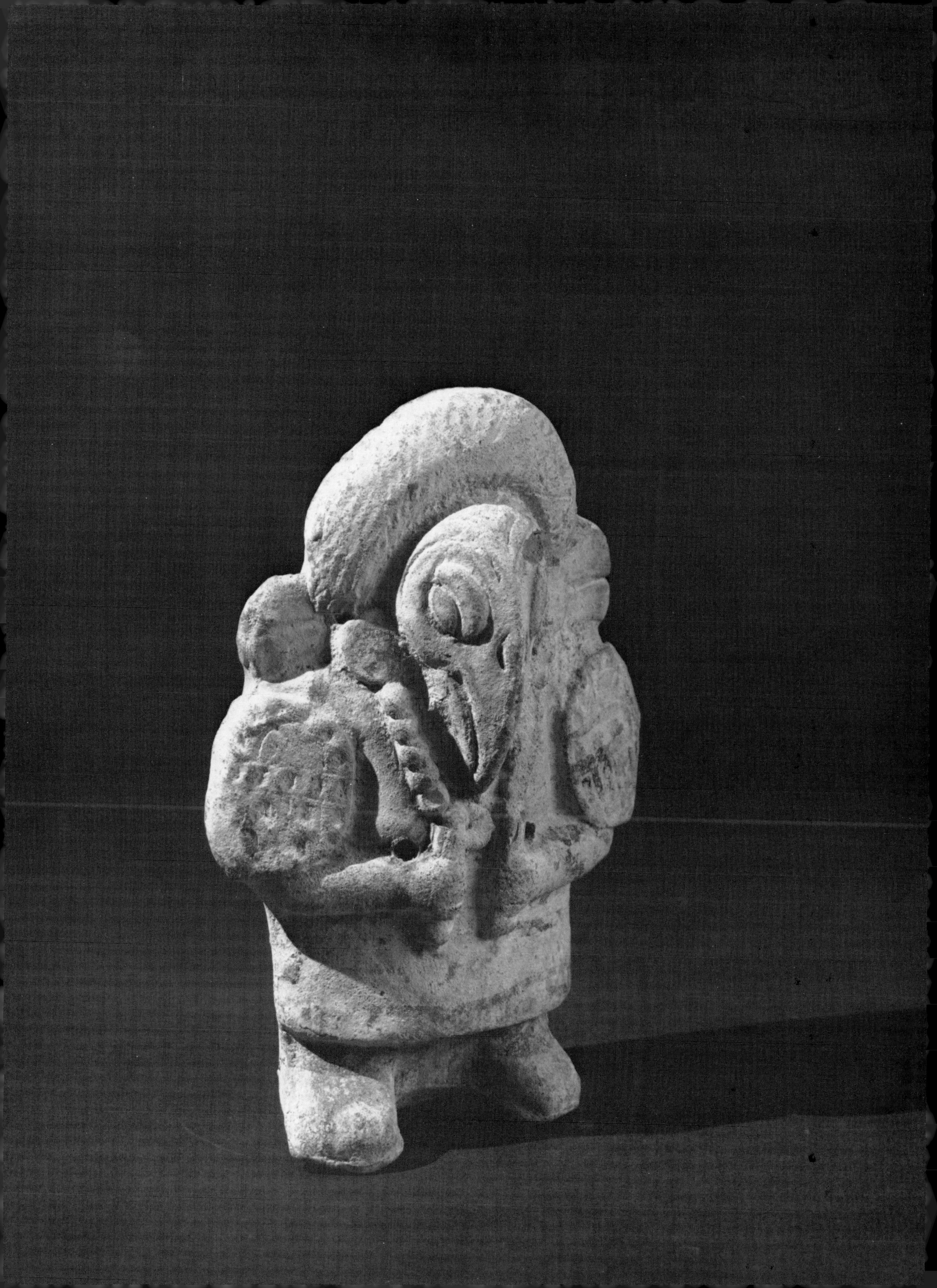

(Fig. 6). The prized blue-green jade in itself denoted vegetation, although the hummingbird's role as a pollinator of plants endowed it with fertility symbolism. Current native American lore highlights both the hummingbird's amorous and military prowess (Benson 1988a). Portions of real hummingbirds are carried as a talisman or eaten to promote courage as well as to infuse new sexual vigor.[6]

PARROT — PAPAGAYO
toznene; ix kan
SCARLET MACAW — GUACAMAYO
alo; kota, mo

The brilliant plumage of parrots and macaws and their ability to fly well above the tree line, also made these birds ideal incarnations of the sky deities. They were sacred to the Cora Indians who considered parrots as Venus, the Morning Star, and, because of their predominently green and yellow feathers, associated them with sprouting corn.

Precolumbian cultures may have likewise endowed parrots with supernatural powers, perhaps those of a celestial origin. Parrot effigies in earthenware appear early in the Preclassic (Fig. 7) and comprise a large number of mortuary goods found in the tombs of West Mexico. Single, double, and triple parrot forms from the Colima culture were hand modeled in clay, painted in two tone slips, and highly polished or burnished (Fig. 8). That the parrot was also adopted as an insignia by the elite of ancient Colima is evident in the parrot supports given *reclinatorios,* head and back rests used by seated figures (Fig. 9).[7]

Both parrots and macaws were prized by all Meso-american peoples for their feathers. Thousands of tiny feathers elaborated costume and dress elements such as capes and headdresses for individuals of status. These birds were among the tropical species demanded as tribute to help meet the demands of the Aztec feather-working industry.

DEER — VENADO
macatl; keh

The nimble speed and bounding grace of deer only partially suggest their fire and stellar symbolism and their recurring role in creation myths. As one of the most valued sources of meat, deer were also featured in ancient fertility rites. Several types of deer were and are still hunted in Mesoamerica, but the most common is the Whitetail deer (Odocoileus virginianus). When frightened, these deer raise their tails like white warning flags. Nomadic peoples who relied most heavily on deer, endowed their super-naturals with deer attributes. For the Chichimecs of north central Mexico, a deer was the disguise of the wife of Mixcoatl-Camaxtli, a god of hunters and warriors affiliated with the planet Venus. Hunting rituals of great time depth were incorporated into the Aztec ceremony that honored Mixcoatl-Camaxtli. Deer were ritually hunted and sacrificial victims were bound hand and foot like captured

deer (Nicholson 1971a:426). Likewise, the deer was celebrated by hunters during the Mayan month of Sip, when each dancer held a deer skull painted blue (Aguilera 1985:26).

The deer is a consistent actor in Maya creation myths. Over time the animal played numerous roles in stories about the formation of the cosmos. Most frequently, the deer is a metaphor for the sun during his courtship of the moon. The sun dons a deer skin disguise as part of his plot to regain his wife, the moon. The sun also orders a deer to conduct a woman, perhaps the moon, to the sky in a Kekchi myth. For the Cakchiquel, two deer draw the sun across the sky. It is also believed that the deer created the genitalia of the Moon goddess with a blow from his cloven hoof. Among the Mopan and Kekchi, the hoof prints of the deer make vulva-like tracks (Fig. 11).

For the Huichols of Western Mexico, the deer serves as a symbol of the hunt and as a metaphor for both corn and the peyote plant. Peyote or *hikuli* is a hallucinogenic cactus and the Huichols' most important ritual substance (Fig. 10). Huichols believe that corn was once deer and the peyote plant first appeared as a giant deer. The deer's blood is shed ritually to insure good crops and a broth of deer and peyote are consumed to make corn grow. A similar belief exists among several highland Guatemalan peoples, where deer blood is smeared on the mouth of deity sculptures in order to promote the growth of maize.[8]

RABBIT — CONEJO
tochtli; tuul; tan kah

In Mesoamerica it is a rabbit-in-the-moon that is seen in the moon's shadows. For the sixteenth-century Aztecs, a full white moon had ''a little rabbit stretched across his face'' (Sahagún 1950–1982, VII:3) (illus. e). According to the Aztec creation myth, the gods struck the moon's face with a rabbit to dim its lustre rather than compete with the sun's brilliance. Although this is a Nahuatl myth, the rabbit as a metaphor for the moon is found in both Classic Maya art and contemporary Mayan mythology.

PAIR OF JADE HUMMINGBIRDS   6
Olmec. Guerrero? Gulf Coast? Mexico
Mid Preclassic: c. 900–500 B.C.
Blue-green jade. L: 3¾"
Private collection

The sophisticated design and skill of Olmec lapidaries is shown in these carved jade hummingbirds. With slits across their upper backs, these birds may have dangled from the headband of an Olmec noble (David Joralemon, personal communication, 1990).

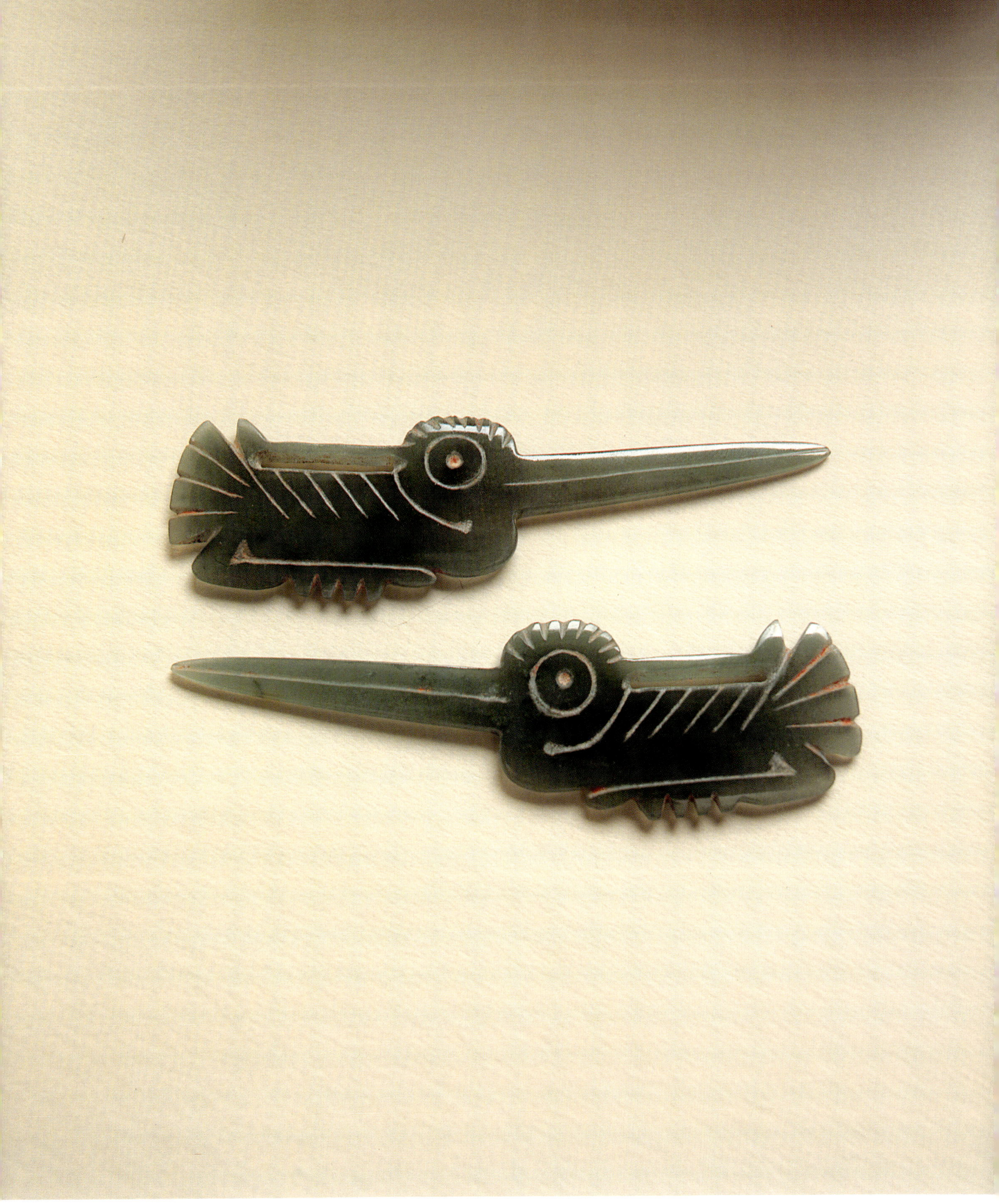

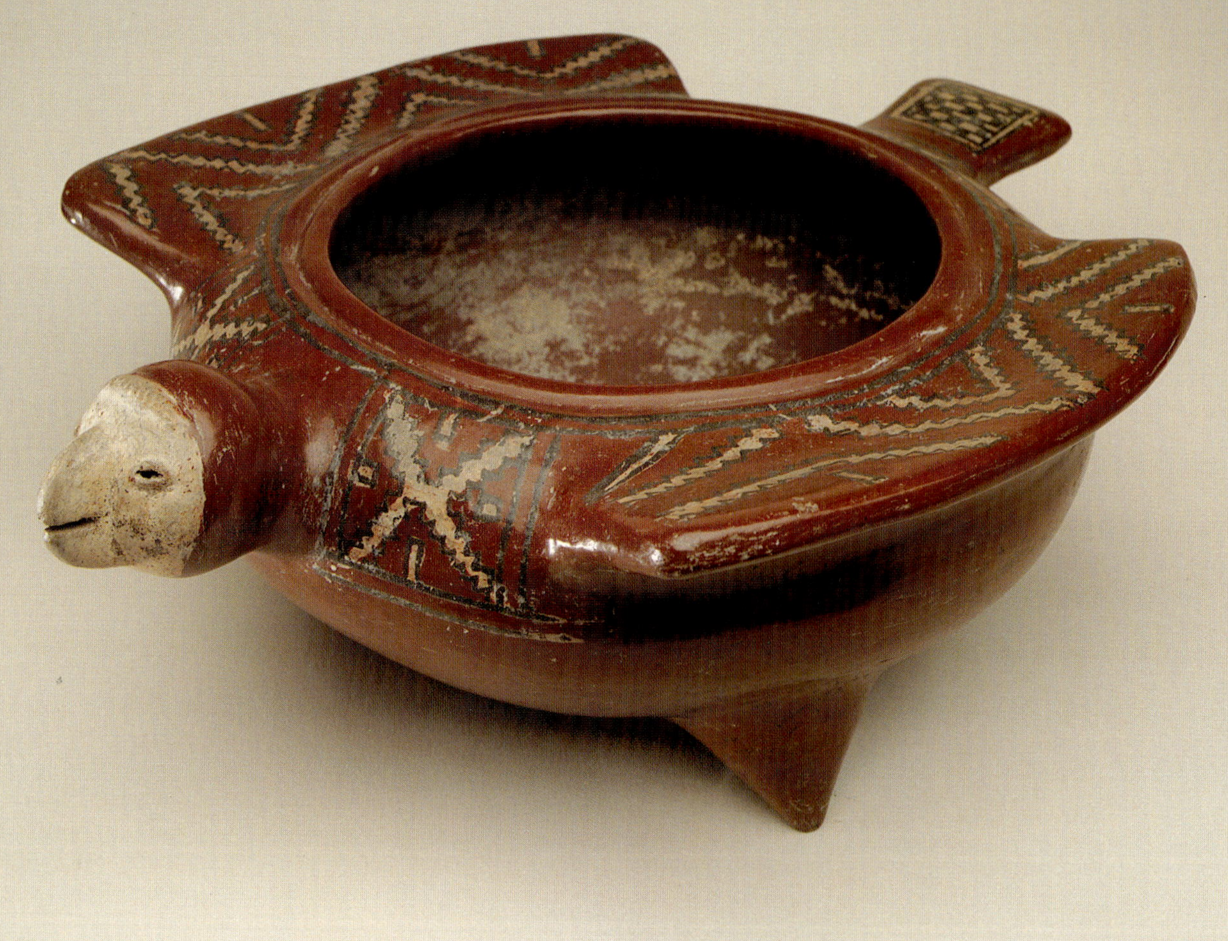

PARROT EFFIGY BOWL 7

Chupicuaro, Guanajuato
Mid Preclassic: c. 900–500 B.C.
Burnished clay with red and black paint. H: 5⅛″

UCLA Fowler Museum of Cultural History, gift of Miss
Natalie Wood 68W.00140

**GLOBULAR VESSEL WITH
PARROT-SHAPED FEET** 8

Colima, West Mexico
Late Preclassic–Early Classic: 200 B.C.–A.D. 300
Earthenware with burnished red-orange slip. H: 10″
Private collection

This pumpkin-like form is often called a ''parrot pot.''
Green and yellow parrots, in addition to having strong
solar symbolism, were also associated with corn and the
earth's bounty.

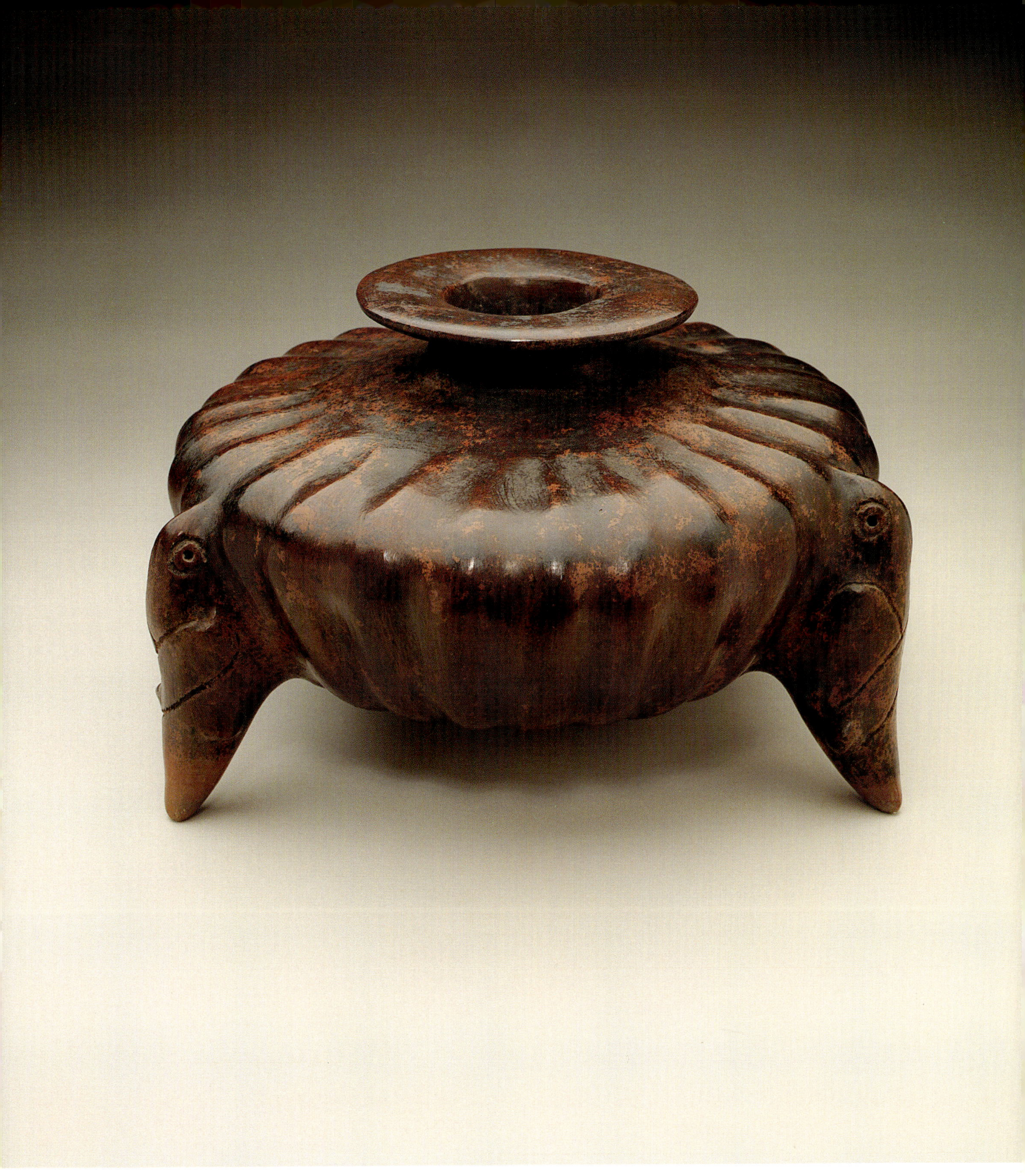

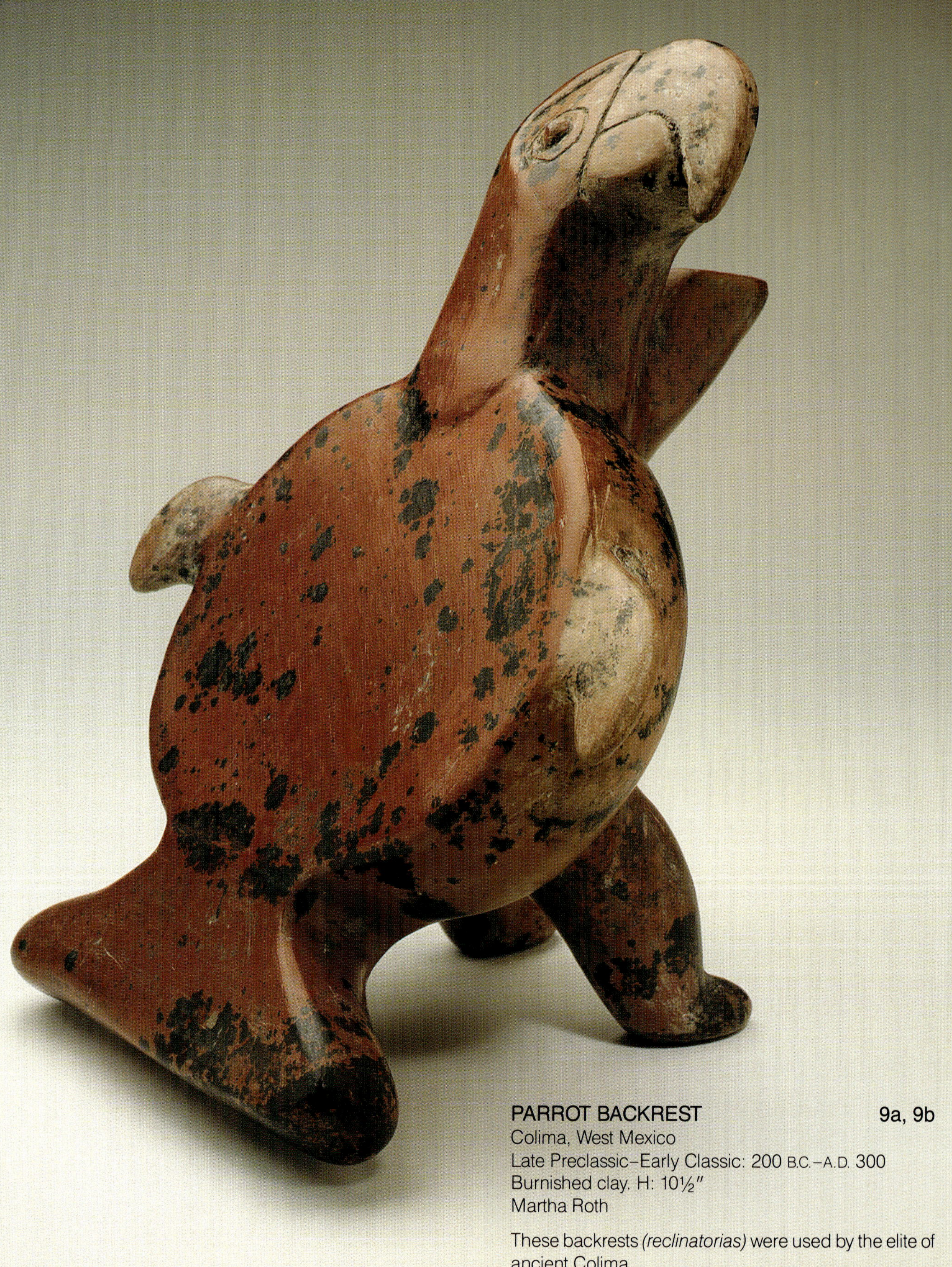

PARROT BACKREST　　　　　　　　　　　　9a, 9b
Colima, West Mexico
Late Preclassic–Early Classic: 200 B.C.–A.D. 300
Burnished clay. H: 10½″
Martha Roth

These backrests (*reclinatorias*) were used by the elite of
ancient Colima.

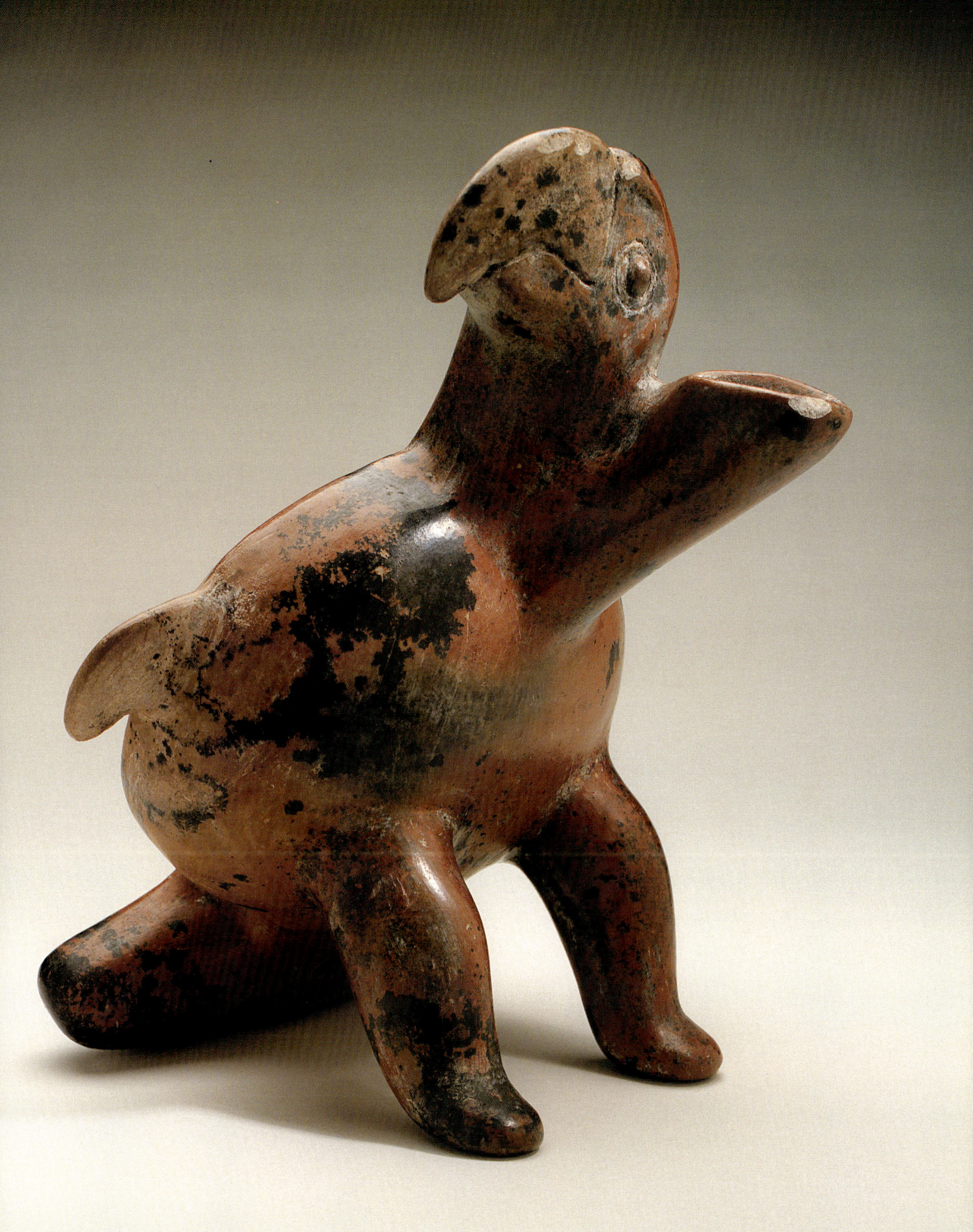

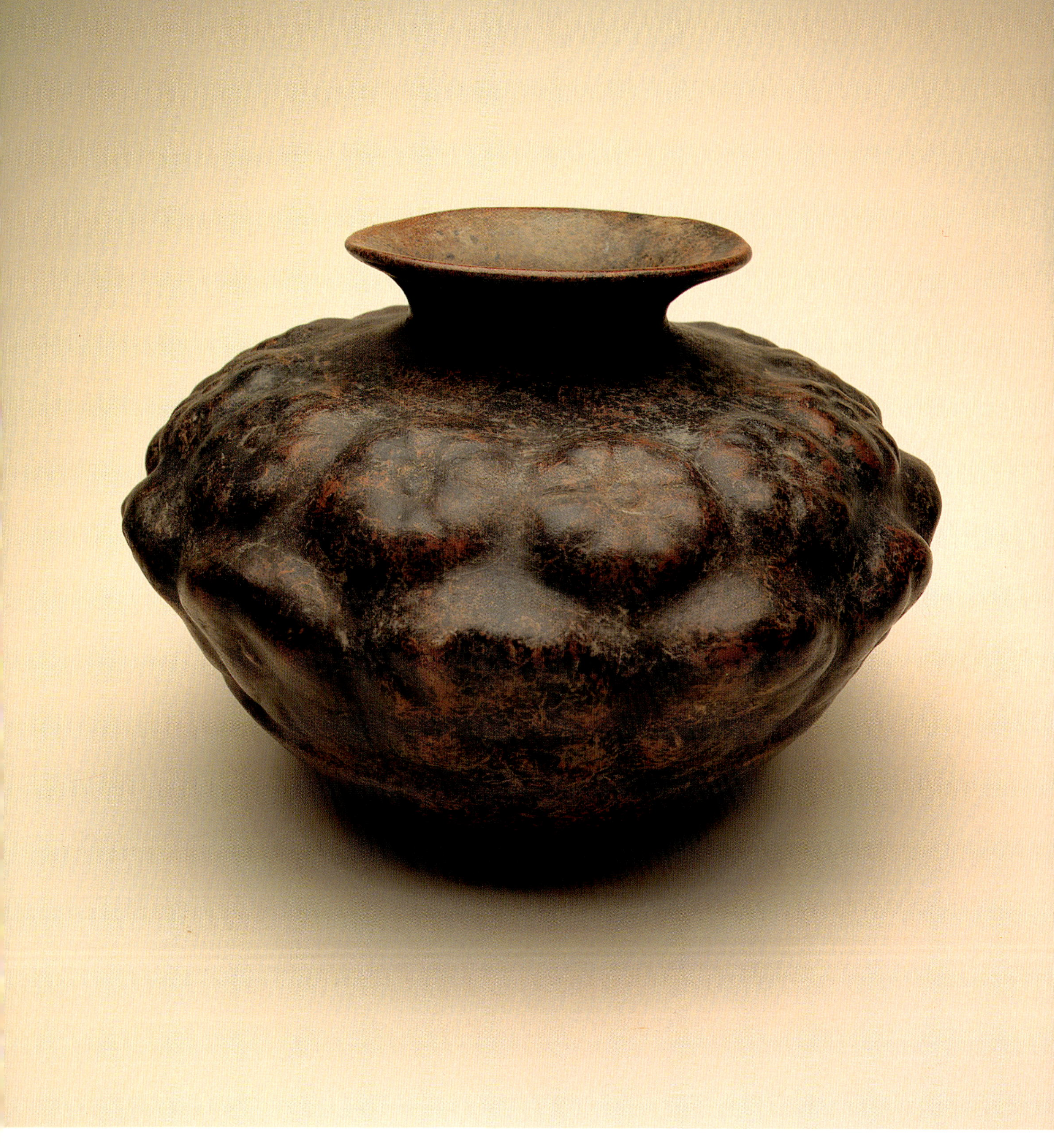

VESSEL WITH PEYOTE   10
Colima, West Mexico
Late Preclassic–Early Classic: 200 B.C.–A.D. 300
Terracotta with burnished dark red slip. H: 8″
Private collection

DEER WITH WOMAN ON PLATE   11
Maya. Petén, Guatemala
Late Classic: c. A.D. 600–900
Polychrome black and red on buff terracotta.
D: 17¼″
Private collection

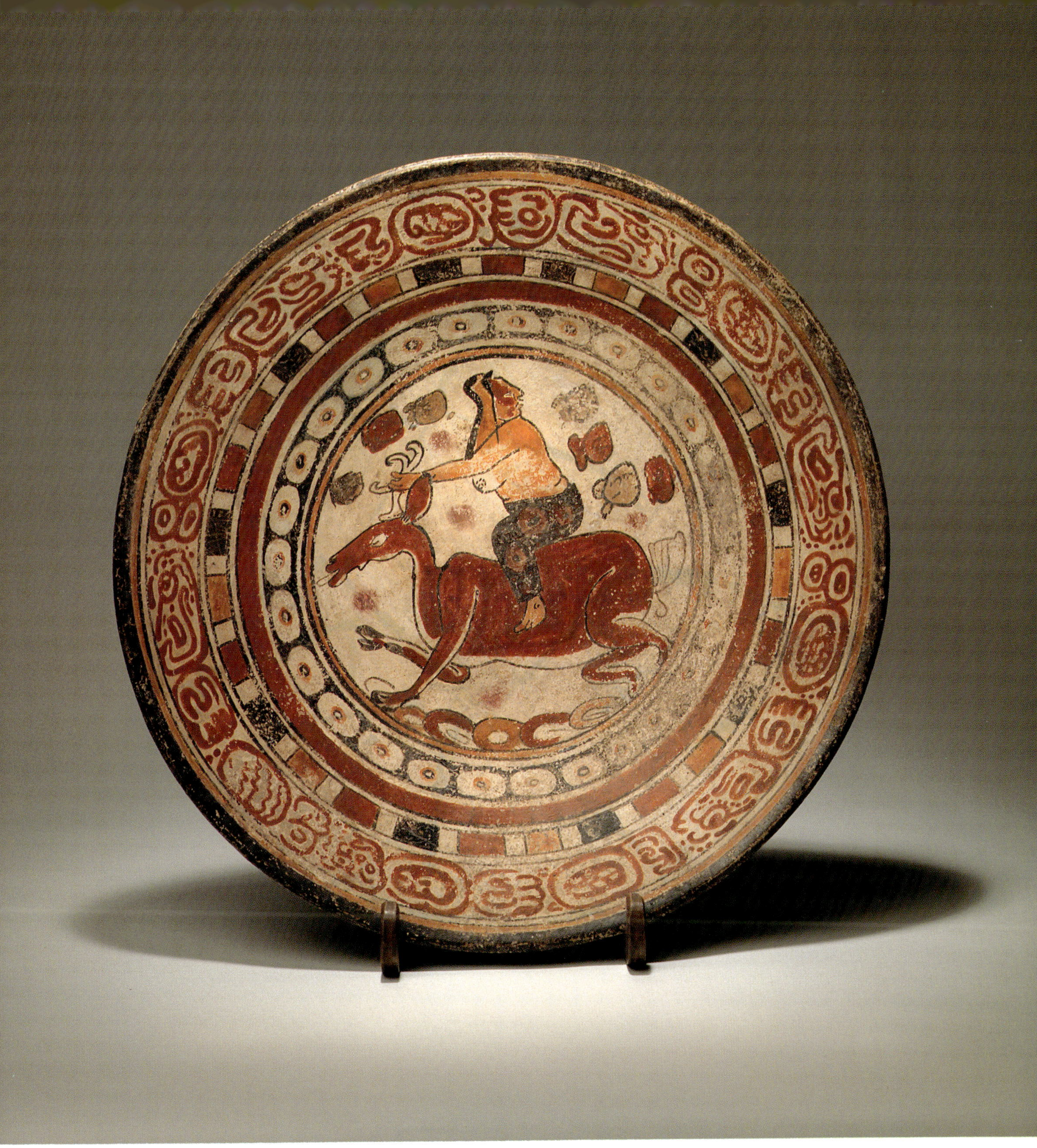

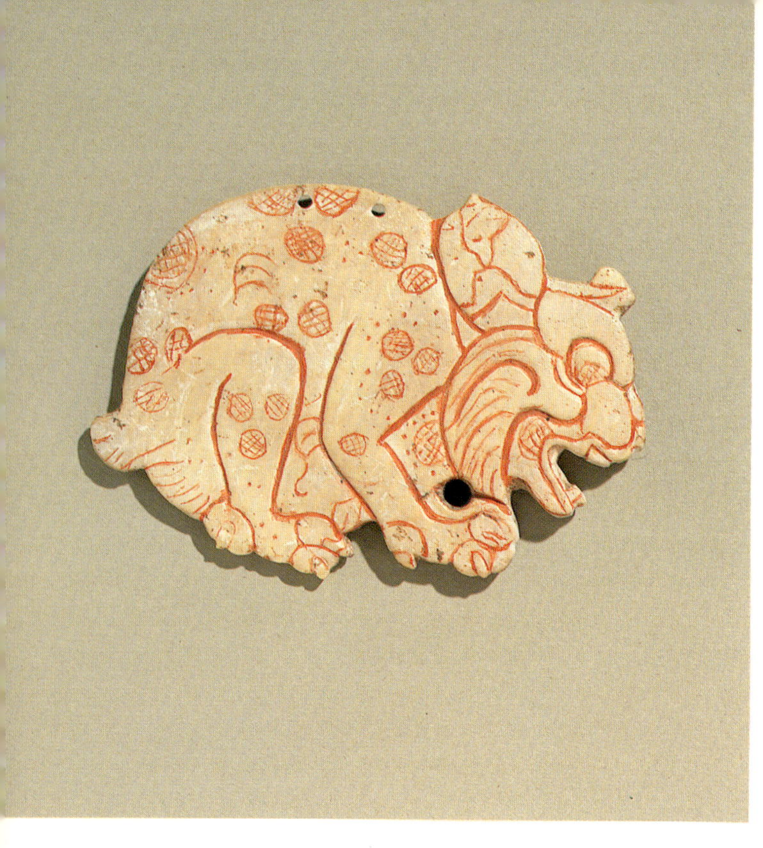

Ixchel, the youthful lunar goddess of the Maya pantheon, could be represented seated on a crescent moon and holding a rabbit (illus. d). Although the Maya moon goddess was an industrious weaver, her other persona was the sexual woman, erotic and fickle, ever-changing like the moon. Here too the rabbit was an appropriate symbol; its vast appetite and proclivity to reproduce from four to seven litters a year, made it the perfect metaphor for fecundity. Rabbits, like deer, have similarly dual associations with both the night sky and the earth's fertility (Fig. 12).

The Aztecs associated the rabbit not only with earth goddesses who controlled vegetation, but specifically with the deities who governed the native alcoholic beverage. This beverage is called *pulque (octli),* fermented from the juice of the maguey plant, and still the most popular drink in rural Mexico. "Two rabbit" was the patron deity of *pulque,* but the entire complex of deities was called the Centzon Totochin (400 or innumerable rabbits), a name that may have designated over-indulgence or excess.

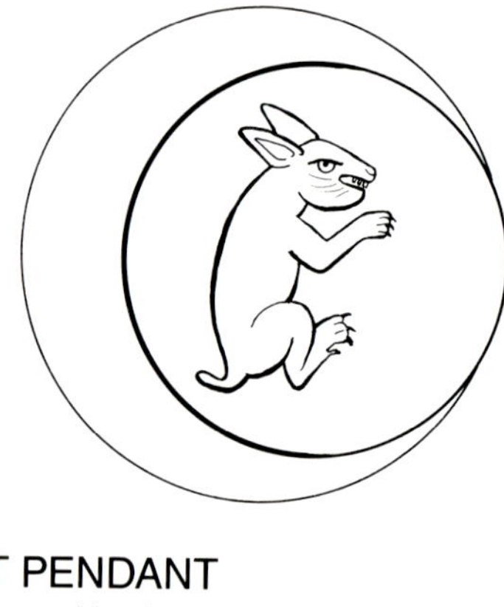

RABBIT PENDANT 12
Maya. Central lowlands, Guatemala.
Mid classic: c. A.D. 550–800
Shell with cinnabar. H: 2″
Carroll collection

As metaphor for the moon, the rabbit is also a symbol of the night sky and earth's fecundity. This Maya rabbit carved out of shell, is a venerable, bearded animal affiliated with the jaguar who reigned over the underworld. The jaguar's spotted coat and highlighted with red cinnabar.

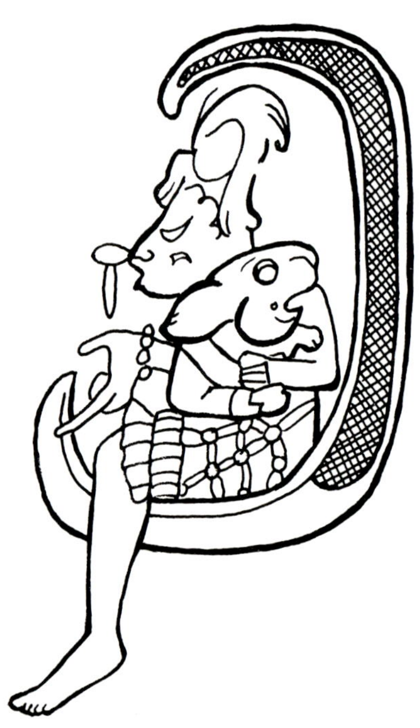

FIGURE WITH RABBIT, 13
WHISTLE AND RATTLE
Maya. Jaina, Campeche
Late Classic: A.D. 550–800   Private collection
Buff clay with red, blue, and black paint. H: 5½″

RABBIT-IN-THE-MOON f
Adapted from *Florentine Codex* Bk. VII: fol. 2r.

THE MOON GODDESS HOLDING RABBIT e
Detail from Late Classic Maya vessel (From Schele and Miller 1986: Fig. 49)

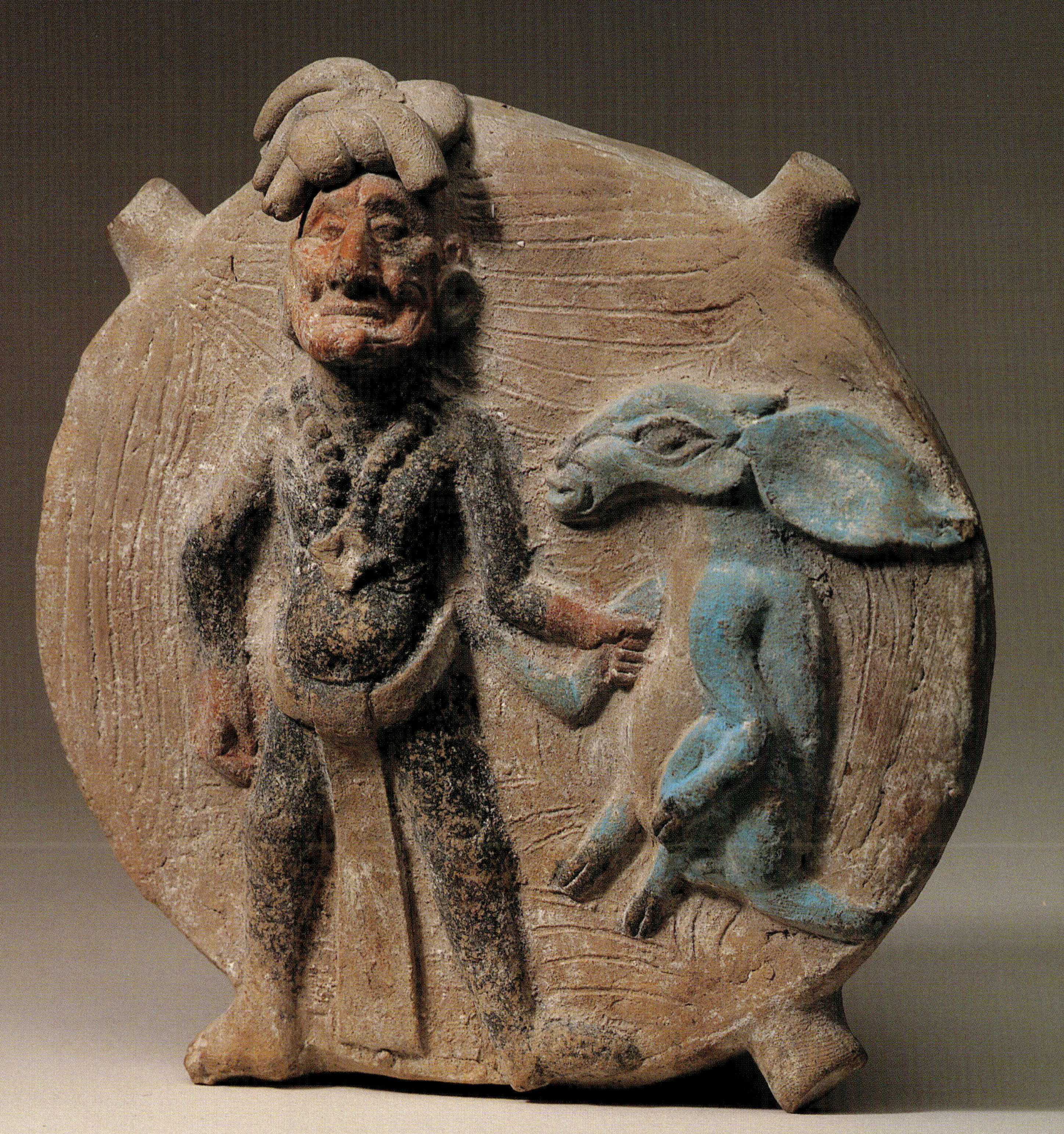

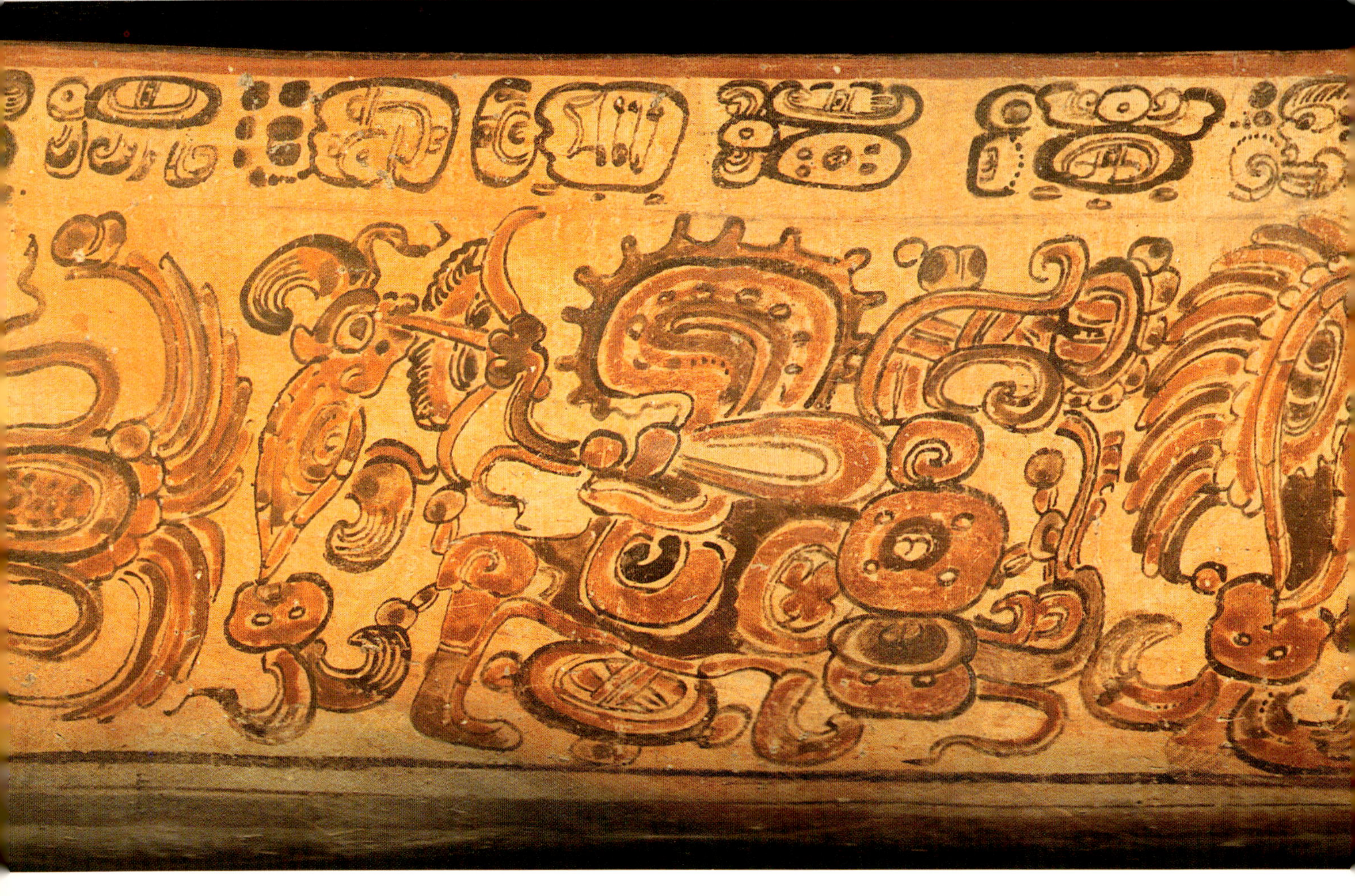

VESSEL WITH PRINCIPAL BIRD DEITY 14

Maya. Tikal, Petén, Guatemala
Mid Classic: A.D. 500–600
Polychrome clay vessel. H: 8″
Stendahl Galleries
Rollout Photograph by Justin Kerr 1990, File #4562

The head and full-figure bird monsters on this vessel represent the Maya Principal Bird Deity. This bird-serpent is now identified by some scholars with "Seven macaw" or Vucub caquix from the Popol Vuh. Characteristics of Vucub caquix include a macaw's turned down beak, twin beaded nose plugs, and a *kin* or sun sign behind the bird's eye, to declare his solar pretensions. The rim text reveals that this clay vessel once held chocolate and probably belonged to a sixth-century Maya Lord of Naranjo (Donald Hales, personal communication, 1990).

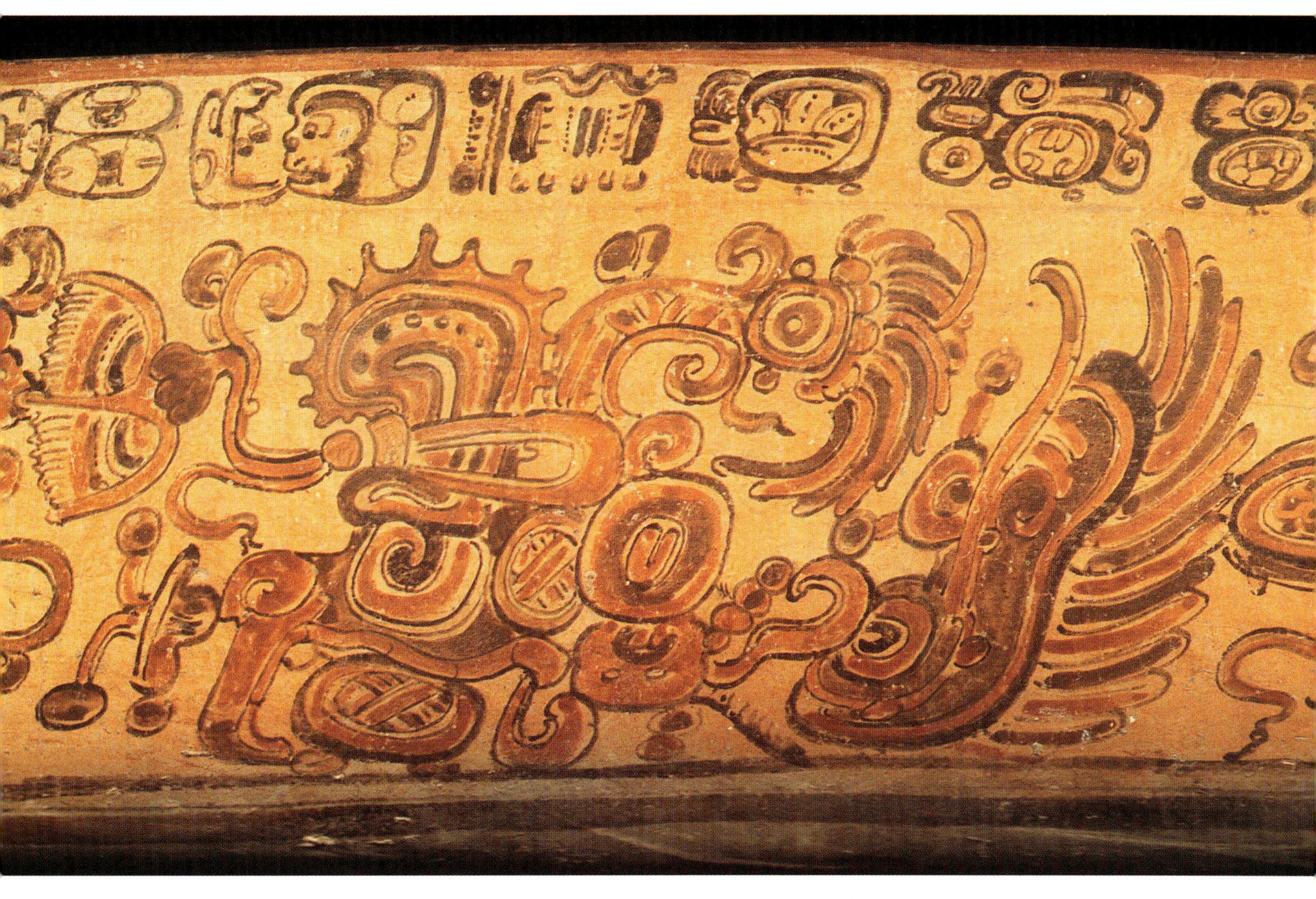

SKY TO EARTH: SERPENT BIRD AND SERPENTS

Serpents occupy a prominent place in Precolumbian belief systems, an importance that is manifest in a wealth of images. Varied in form and meaning, the serpent inspired sky and earth imagery. The double-headed serpent, like the fire serpent, is a celestial entity; the famed plumed serpent and serpents in general, are symbols of the earth's regenerative powers.

SERPENT BIRD

A bird with serpent features, most frequently called the "Principal Bird Deity," evolved into one of the most important Classic Maya deities. The inverse creature, a serpent with feathers, Quetzalcoatl or Kukulcan, played a pivotal role among the central Mexicans and Post-classic Maya. The fusion of snake with bird is an example of the Mesoamerican propensity to unite seemingly contradictory elements. These dual-natured animals were accorded dominion over the heavens and the earth, powers that were adopted by the ruling hierarchy. Although snakes and birds are predatory enemies from the opposing realms of air and earth, they share a common evolutionary origin (Clancy et al. 1985:197). They are both "twice-born," as eggs first and then as divergent creatures. Reptile scales are the biological prototype of feathers; conversely, dense, gleaming feathering, such as the hummingbird's, can look like fish scales. As a further analogy, snakes, like toads, shed their skin and birds moult. This process made the reptile-bird creature a powerful symbol of transformation.

PRINCIPAL BIRD DEITY — VUCUB CAQUIX

The Principal Bird deity is one of the most challenging celestial birds in the profuse world of Mayan supernaturals. As first described by Bardawil (1976), the Principal Bird deity combined a bird's head and an outstretched "serpent wing." The feathers of the serpent wing are secured to the mouth of a stylized serpent head with a long extended upper lip. This bird monster can be traced back to narrative scenes carved on Izapan stone stelae of the late Preclassic. The Classic Maya at Palenque revered it as their celestial bird on top of the world tree, wearing a mirror device and god markings. The Principal Bird deity was closely associated with Maya rulership, and the serpent wing was worn in royal headdresses.

Several specific identifications of the Principal Bird deity have been made, most recently with the Harpy eagle and with "Seven Macaw" of the Popol Vuh. "Seven macaw" or Vucub caquix is described with red feathers and a white beak, like a Scarlet macaw (Ara macao). In the Quiché narrative, Vucub caquix presided over a previous cosmic epoch and assumed omniscient powers, claiming, "I am their sun and I am their light" (Tedlock 1985:86). The hero twins succeeded in defeating the arrogant Vucub caquix and breaking his jaw. It is this incident that gave the celestial bird his latch-like snout, like a macaw's overhanging upper mandible. Vucub caquix was clearly an animal with dual powers; his image is simultaneously associated with the kin or sun sign as well as the akbal or earth symbol.[9]

The permutations of Mayan supernaturals are apparent in florid images of the Principal Bird Deity. Both the bird monster's ornate head as well as a full figure rendering with a serpent wing and tail can appear painted on a single Maya vessel (Fig. 14). A reading of the hieroglyphic text around the rim of this vessel, called the Primary Standard Sequence, reveals not an amplification of the image, as might be anticipated, nor a ritual chant, as was first hypothesized. Rather, the rim text is generally a prosaic account of the vessel's medium, contents, and owner (Stuart 1989). In this case, a loose translation of the text might read as follows: "What is written on the vessel of [or by] the Lord of Naranjo is about his food, cacao [chocolate]." The Lord of Naranjo reigned for about forty years, from A.D. 554–593 (Donald Hales, personal communication, 1990).

Mayan clay vessels such as this one functioned in both everyday and ritual contexts during a man's lifetime. They held food stuffs, most commonly the important *atole* or corn gruel and cacao as a chocolate drink. However, when buried with the dead, was the food intended to be used by the deceased in the afterlife or was it meant as godly food? The question as yet unanswered, but the chocolate beverage may well have been an offering for the gods. Justin Kerr (personal communication, 1990) relates that he observed a private offering to a stone monument in rural Guatemala several years ago. "Aside from the copious amounts of *copal* [incense] that were being burnt, a bar of chocolate had been placed in the mouth of the idol. It is exciting to think that a sentence on an early classic vessel is still a viable message in their world."

COILED FEATHERED SERPENT   15
Aztec. Central Mexico
Late Postclassic: c. 1400–1519
Stone (andesite?). H: 10″
Collection of Sanford M. Gross

The serpent's scales were converted into the long, flowing feathers of the Quetzal bird in this superb Aztec example of the mythical Quetzalcoatl (Plumed serpent). This fusion of bird and snake combined the forces of sky and earth into one powerful supernatural.

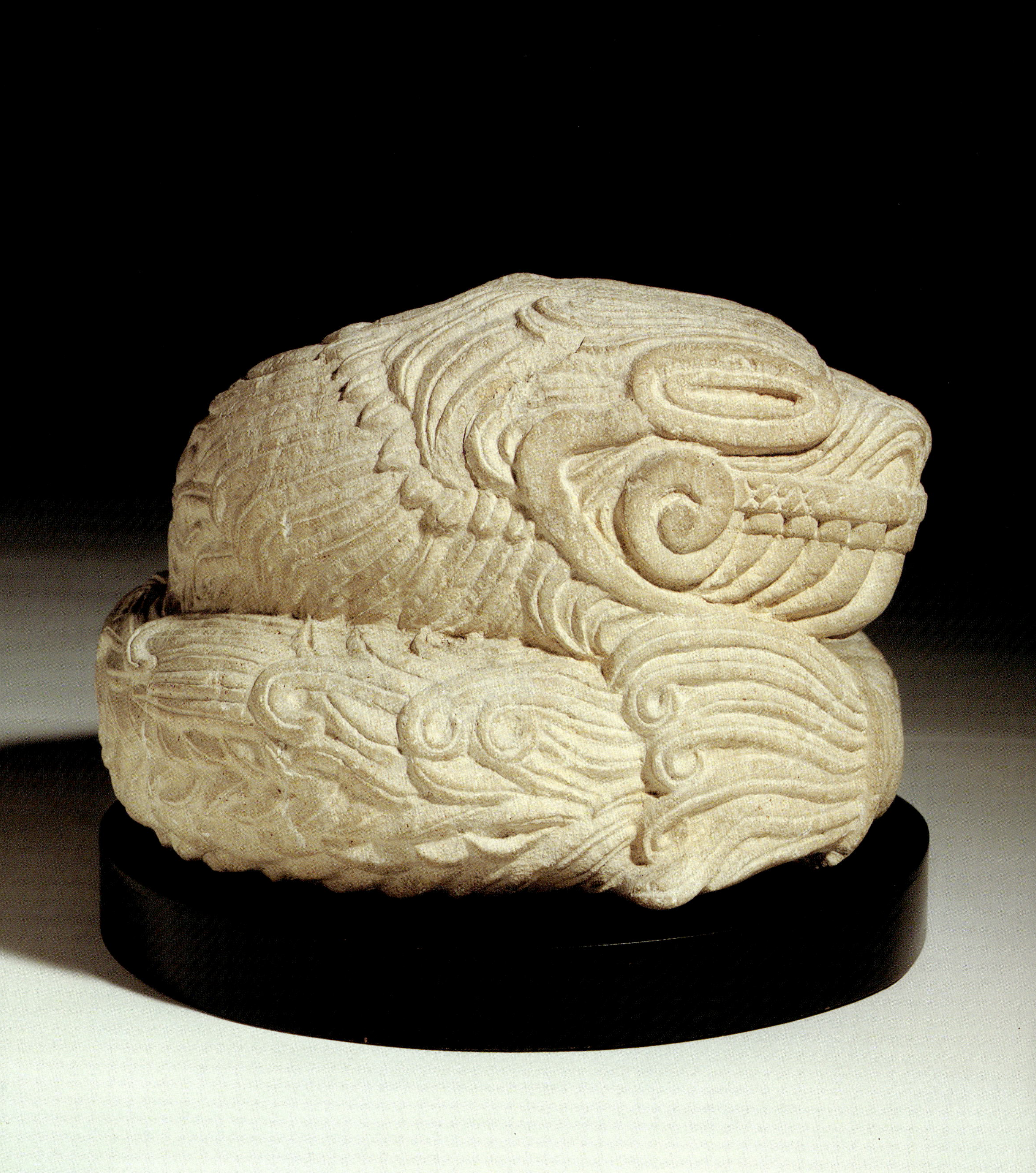

THE FEATHERED SERPENT — QUETZALCOATL

The feathered serpent, known by its familiar name of Quetzalcoatl, has long intrigued the popular imagination. A serpent with a feathered head or body is found in the art of the earliest Olmec civilization, the Classic Teotihuacan culture, and became pan-Mesoamerican in the Postclassic (Fig. 15). The plumed serpent merges the most important of serpents, the rattlesnake, with the splendid quetzal bird (Pharomachrus mocinno). As a coveted tribute item, the quetzal was ardently sought in its cloud forest home. The magnificent train of the male quetzal is made up of two tail feathers or coverts that overlay its shorter tail by as much as 20 inches. These luxuriant feathers were considered rare and sacred because of their very color, a vivid green with a blue sheen. Blue-green materials such as feathers and jade, were precious to all Mesoamerican high cultures as they connoted water, fertility, and life itself.

The use of the quetzal bird underscored the divine descent and prominence of the supernatural Quetzalcoatl. In opposition to Quetzalcoatl who symbolized the verdent earth, the Aztecs juxtaposed Xiuhcoatl (Turquoise serpent). Xiuhcoatl was a fire serpent who swallowed celestial bodies at one end and then expelled them at the other. As complementary opposites, Xiuhcoatl and Quetzalcoatl were paired metaphors for the eternal cycle of the hot dry season and the lush, rainy season (Pasztory 1983:234).

A feathered serpent may have been associated with Olmec elite figures, probably priests (Joralemon 1988:11). A priestly ancestral figure was also the hero of the best known Quetzalcoatl saga. According to Aztec myth-history, a shadowy tenth-century Toltec leader named Topiltzin Quetzalcoatl, promised to return in the year One Reed *(ce acatl)* after being driven from his native city. By native count, 1519 was the year One Reed. It was in 1519 that the Spanish conqueror, Hernan Cortes, unwittingly fulfilled Quetzalcoatl's prophecy by landing in Vera Cruz. The wily Cortes turned this coincidence to his advantage in the conquest of the Aztec empire. The Postclassic Mayans adopted the cult to Quetzalcoatl but called him by his Yucatec name, Kukulcan; feathered serpent columns and reliefs adorn many buildings at the site of Chichen Itza in Yucatan.

TWO-HEADED SERPENT

A two-headed serpent arched over the earth or ringed the cosmic seas. The body of the two-headed serpent was often conceived of as a sky band, serving as a passageway for the planets and stars. Two-headed serpents surrounded mask images representing the sun and Venus on Preclassic Mayan temple façades at Cerros, Belize. The double-headed serpent appeared frequently on thrones, costumes, and sculpture as a border that delimited the Maya world (Carlson 1988a).

Among the Classic Maya, a bicephalic serpent or dragon became a signifier of kingly rank on headdresses, belts, and the ceremonial double-headed serpent bar. When held by the king, the ceremonial bar announced he quite literally supported the sky in his arms. In Yucatec Maya, the same word, *caan (k'an)* is used for "serpent" and "sky."

The appearance of the double-headed serpent on Colima, West Coast, pottery may depict a similar celestial creature (Fig. 16). For the nineteenth-century Huichol Indians, the sea (i.e., the world) was surrounded by a great devouring serpent with two heads.[10]

SERPENT — SERPIENTE   *coatl; caan* or *k'an*

Precolumbian man felt the same ambivalence toward the snake as he did toward the earth — it too was considered beneficial as a source of nourishment and also, potentially deadly. The serpent's unusual methods of locomotion, slithering on the ground without the use of legs and swimming with no fins, put it in a special category. Important too, is the snake's annual shedding of its skin. This capability to rejuvenate itself made it a symbol of renewal and health (Fig. 17). Yet, the serpent could kill by suffocating its prey in the powerful coils of its body or impale a victim and inject venom with its long curved fangs, as does the venomous rattlesnake (Fig. 18).

In central Mexico, the sight of some snakes portended evil and danger. However, when an individual was born under the calendrical sign Snake, the priestly predictions were largely favorable. "Seven serpent" represented sustenance and the day sign "one serpent" brought renown and honor.

As positive symbols, serpents were metaphors for rain and blood, equally perceived as life-giving liquids. Associated with a rain cult that had ancient roots in Mesoamerican history, serpents took twhe form of rain deities or served them in one capacity or another. Thompson (1970:265) has suggested that the Maya Chacs, who were gods of rain, thunder, and agriculture, were originally derived from the snake form. In Classic Maya imagery, the deity Chac emerges from the jaws of a serpent. For the present-day Chorti, four giant snakes or

---

URN WITH PLUMED SERPENTS
Mixtec, Puebla, Mexico
Postclassic: A.D. 1200–1500
Polychrome terracotta
Collection of the Denver Art Museum,
Gift of Morton D. May
Photography courtesy of D.A.M.

Chicchans live at the four points of the compass. During the dry season, the Chicchans remain in the sky, however, in the wet season, they reenter springs and rivers. The rain deities of the modern Tzotzil are seated on snakes. Snakes are the servants of the Kekchi lord of rivers.

The Mexican rain god, Tlaloc, was also associated with the serpent; the deity often holds a snake symbolizing lightening and rainstorms. Serpents were important insignia of Aztec earth goddesses, such as Coatlicue (She of the Serpent skirt) whose skirt is woven with intertwined snakes. The most prominent goddess of corn, always shown holding double ears of maize, was named Chicomecoatl or Seven serpent (Fig. 19).[11]

The role of serpent effigies is as problematic as the many other naturalistic animal sculptures executed by Aztec craftsmen. They may have functioned variously as place glyphs and cult objects. The large undulating stone serpents and serpent heads on the Aztec Templo Mayor, for example, identify the pyramid with the mythic Coatepetl (Serpent Mountain), the sacred space where the gods were born (Broda 1987:77). Snake sculptures may have also amplified a certain deity's powers, as do the frog sculptures placed near the shrine to the rain god, Tlaloc, or the squash sculptures in temples dedicated to earth goddesses.

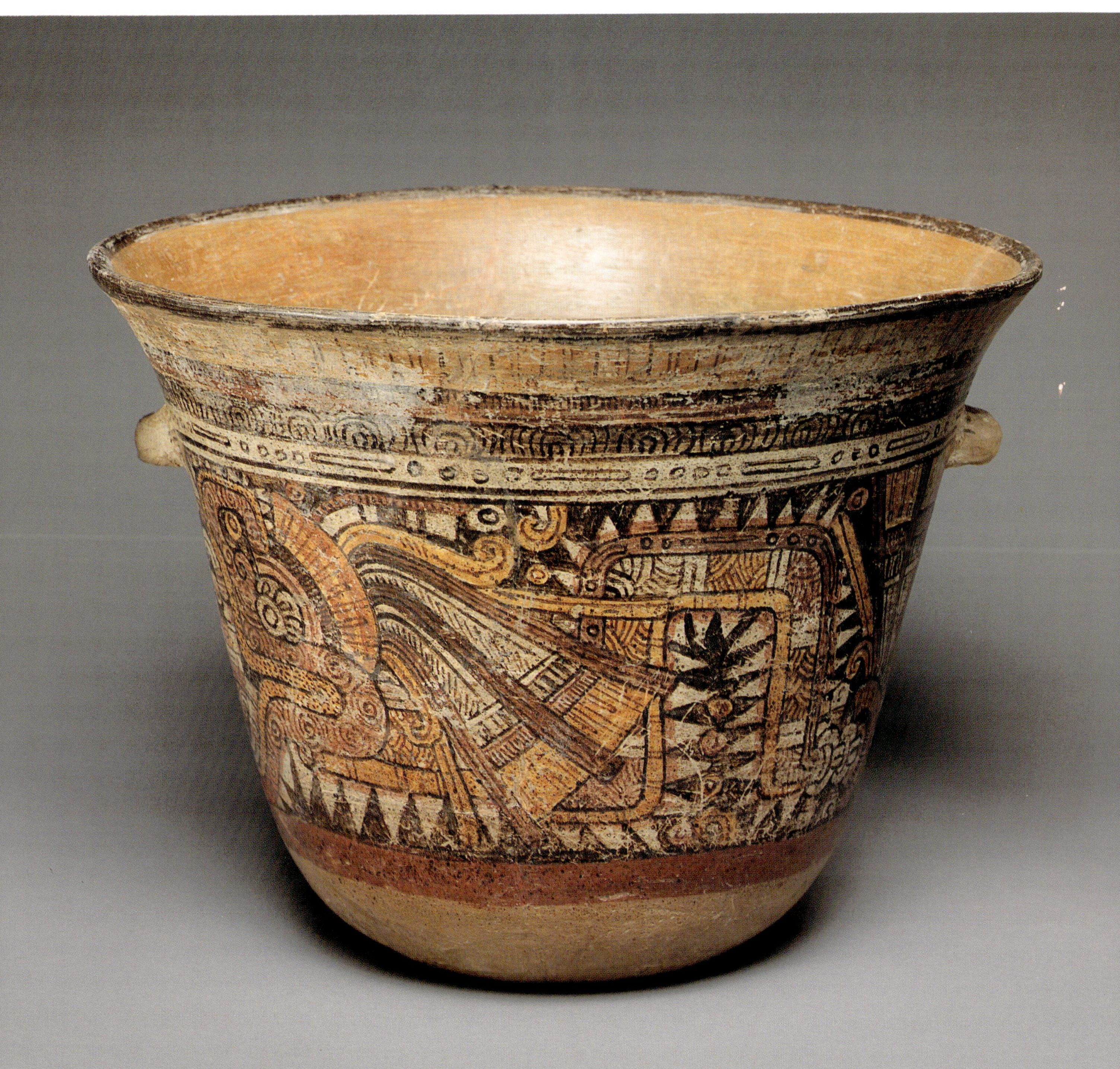

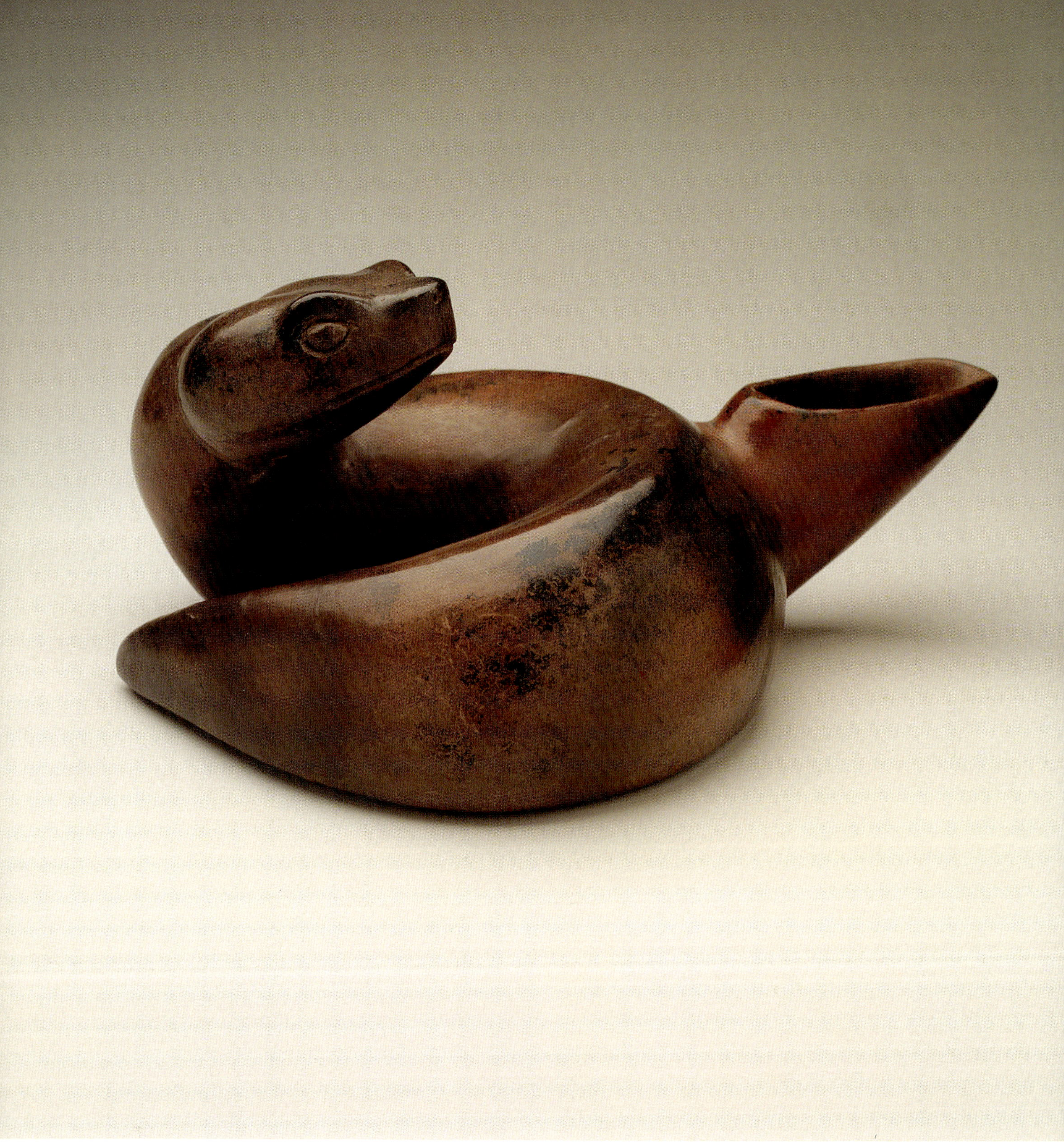

COILED SERPENT EFFIGY   17
Colima, West Mexico
Late Preclassic–Early Classic: c. 200 B.C.–A.D. 300
Burnished red slipped terracotta. H: 10″
Private collection

BOWL ON DOUBLE-HEADED SERPENT   16
Colima, West Mexico
Late Preclassic–Early Classic: c. 200 B.C.–A.D. 300
Earthenware, Monochrome Red. H: 16½″
The Bowers Museum, Santa Ana   F79.82.1
(In Labbe 1982: cat. #10)

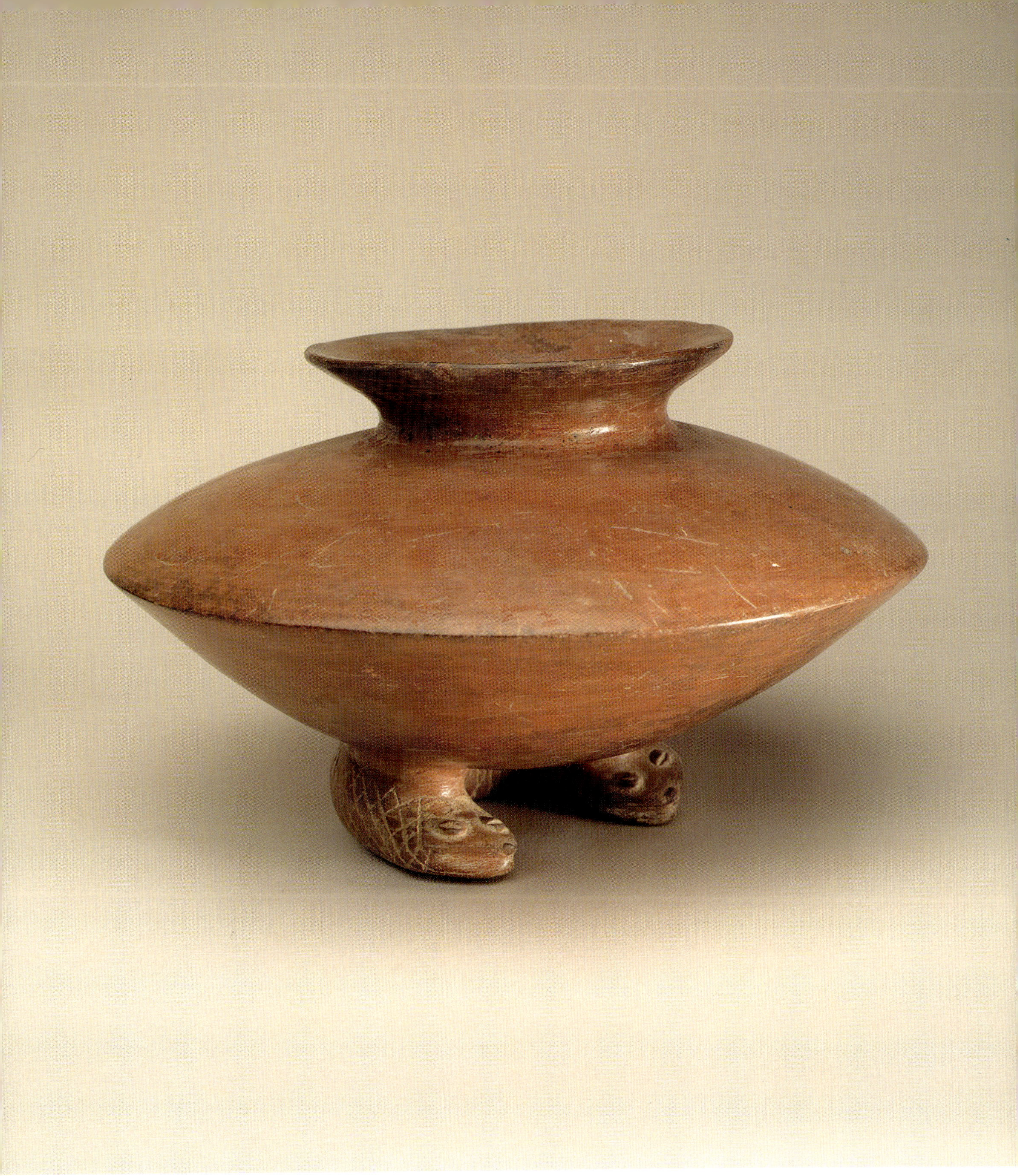

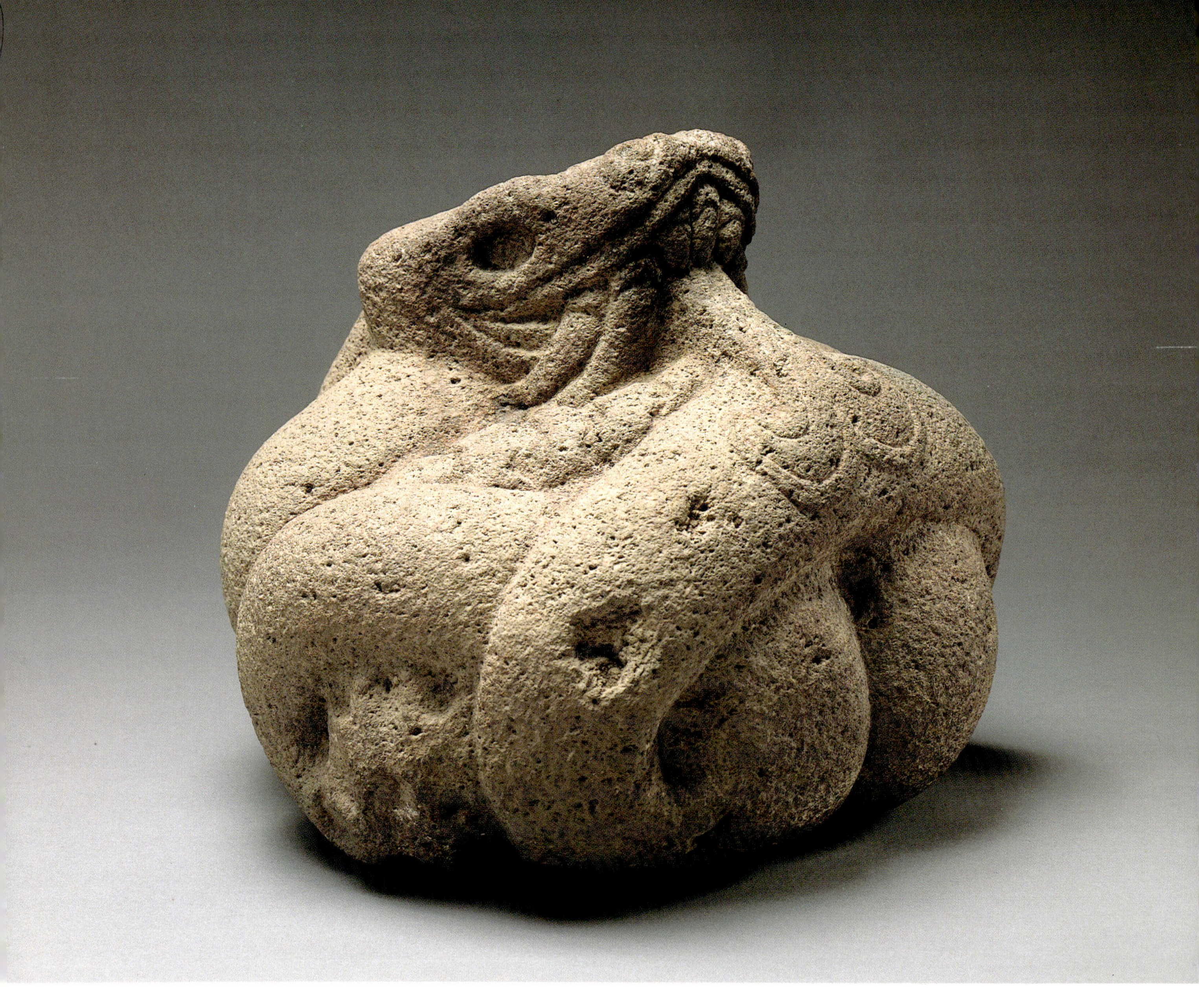

<table>
<tr><td>

KNOTTED AND COILED RATTLESNAKE 18
Aztec. Valley of Mexico, Central Mexico
Late Postclassic: c. A.D. 1450
Stone. H: 10½″
Collection of The Denver Art Museum, museum
purchase with funds from the Burgess Matching Trust
and other donors. 1989.8

</td><td>

CORN GODDESS (Chicomecoatl) 19
Aztec. Valley of Mexico, Central Mexico
Late Postclassic: c. A.D. 1500
Basalt. H: 17⅜″
Collection of The Denver Art Museum, museum
purchase 1957.31

This cult statue is of the Aztec maize or corn goddess
named Chicomecoatl (Seven Serpent). The goddess is
shown holding double ears of corn and she wears a
Temple Headdress, a stone replica of a paper headdress
with rosettes and streamers.

</td></tr>
</table>

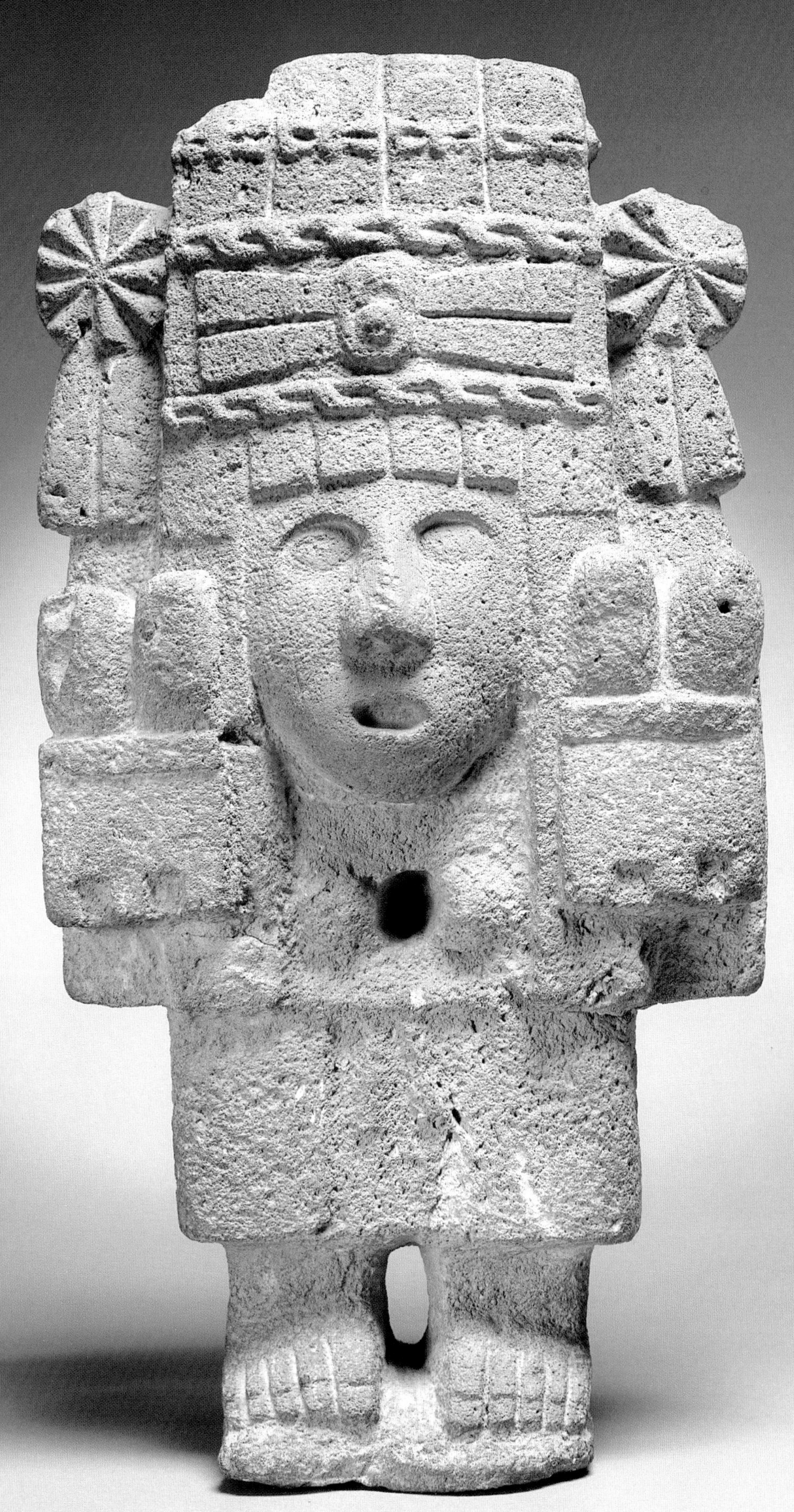

EARTH

When it breaks the water's surface, the dark green and brown shell of a crocodile or turtle resembles an island. A reptile, floating in water or a lily pond, was a graphic metaphor for the earth's surface. One Nahua creation myth recounts that the primeval earth was formed from a reptilian creature; "...from the waters they created an enormous fish known as *cipactli,* which was like an alligator and from this fish the earth was made." (*Historia de los Mexicanos,* in León Portilla 1963:35). In the Maya Popul Vuh, the deity-maker of mountains and earth was a mythological animal, who was part crocodile and part shark. Crocodiles, turtles, and toads recreated cosmic paradigms and carried rich associations of water and fertility. All three are amphibious, with the capability of swimming in water and moving on land. Crocodiles and turtles are protected by bony plates, while toads have a thick almost horny epidermis. Each displays geometric designs on their skins or shells that suggest the regular pattern of tilled fields.

The generous Mesoamerican earth supported diverse plant life (Figs. 21, 22, 23) and an abundance of crops. The three pivotal food staples were maize (corn), beans, and squash, supplemented by sweet potato, chile pepper, and many fruits, including the indigenous tomato and avocado. These foods are curiously absent from Mesoamerican art, with the exception of the large family that includes gourds, pumpkins (Fig. 9), and squash. The dried shells of bottle gourds were, and still are, decorated to serve many purposes, from containers to musical instruments.

CROCODILE — COCODRILLO, CAIMAN, LAGARTO,
cipactli; ain

A crocodile-like creature affiliated with the earth appears as early as the Olmec dragon. Although the Olmec dragon is a composite, mixed with Harpy eagle and jaguar characteristics, its dentition and jaw appear to be that of a cayman (Caiman fuscus) and its gaping mouth a symbol for the cavern leading to the underworld. The three Mesoamerican families in the Crocodilia order, crocodiles, alligators, and the smaller cayman, were often treated similarly in Precolumbian art and myth. Moreover, in some instances the lizard was likewise equated with the crocodile. In fact the Spanish *lagarto,* or large lizard, is the origin for the English word alligator.

A crocodilian earth monster is sometimes depicted giving rise to the World Tree. This image can be charted from Preclassic Izapan sculptural reliefs (illus. g) to illustrations in the Postclassic Mayan codices. Thompson (1970: 212–224) argues that an "Iguana earth crocodile" (Itzam Cab Ain) was the floor or terrestrial sphere of the Maya universe. The *cipactli* (alligator-lizard) was adopted by one of the Aztec creator gods, Cipactonal.

Given its watery habitat in swamps and rivers, the crocodile is also a natural metaphor for fertility. The first day signs in the Mayan and Mexican calenders signifying water and abundance are both related to the crocodile. The crocodile (Crocodylus moreletii or C. acutus) was as important a symbol for the food-producing land among the Classic Maya as was the water lily (Puleston 1977). The pattern on crocodiles' scaly backs appears to replicate the grid design of rectangular raised fields intersected by irrigation canals. On a Copan altar, waterlilies and fish nibble on the extended limbs of crocodile-monster (illus. h).

TURTLE — TORTUGA   *Ayotl; Ak*

While some varieties in the large turtle family are plentiful in Mesoamerica's marshland, other turtles are equipped to navigate long sea voyages. Turtle imagery was used to indicate the watery surface that separated sky from underworld. On an Early Classic Maya bowl from Tikal, a turtle is modeled on the flange and supports a lid capped by the head of a water bird, one of the animals thought to bridge the sky and underworld (Fig. 25). In other Postclassic art works, the turtle suggests the four-part division of the cosmos (Fig. 26).

In addition to their role as earth metaphors, turtles, like crocodiles, had fertility connotations. The turtle was considered an ally of the Maya rain god, Chac. Its carapace was worn by the Mayan bacabs, or avatars of the rain god, as it was by the Aztec deity Tonacatecuhtli, Lord of Sustenance. Crocodile skeletons and turtle shells that accompanied Maya rulers to their graves, may have also signified fertility and new life in the Otherworld.

Splendid terracotta sculptures of turtles are among the many animal effigies created by the Precolumbian artists of West Coast Mexico. In addition to the naturalistic turtle vessels of Colima are bizarre human-turtle combinations. Sculptures that combine the turtle shell with a human head also occurred in Classic Maya, Aztec, and West Coast art. The portrayal of a human head protruding from the turtle's shells suggests themes of ancestry and emergence, including the germination of life from an earth-turtle supernatural.

CROCODILE TREE SURMOUNTED BY BIRD g
Preclassic Izapa Stela 25. (From Hellmuth 1987: Fig. 595)

**CROCODILIAN EARTH MONSTER   h
WITH WATERLILIES & FISH**
Maya stone Relief, Altar T, Copan, Honduras. (From Puleston 1977: Fig. 4)

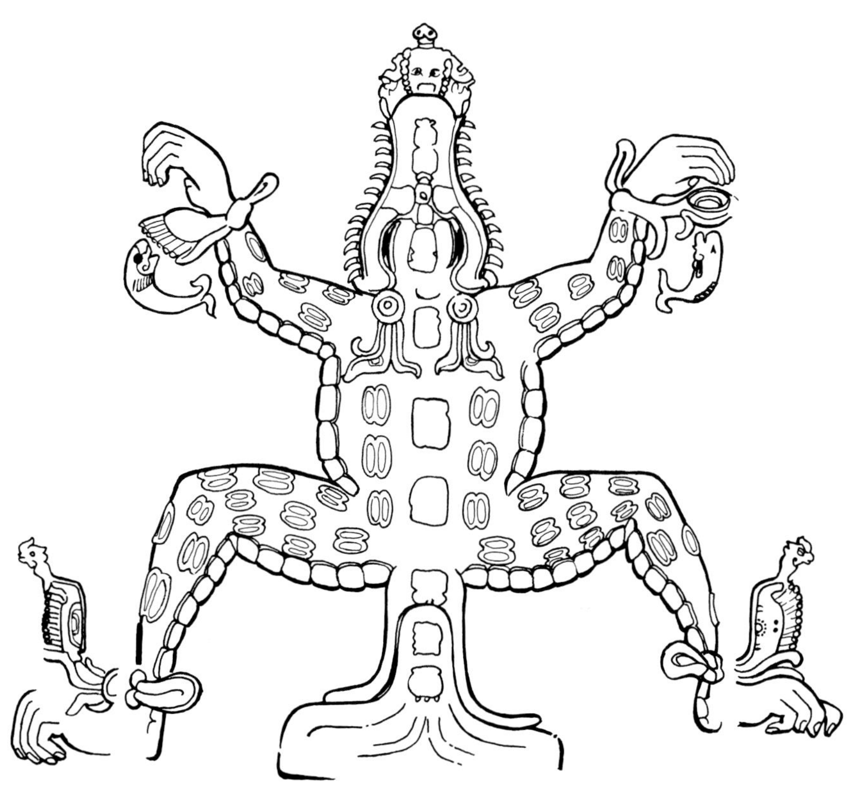

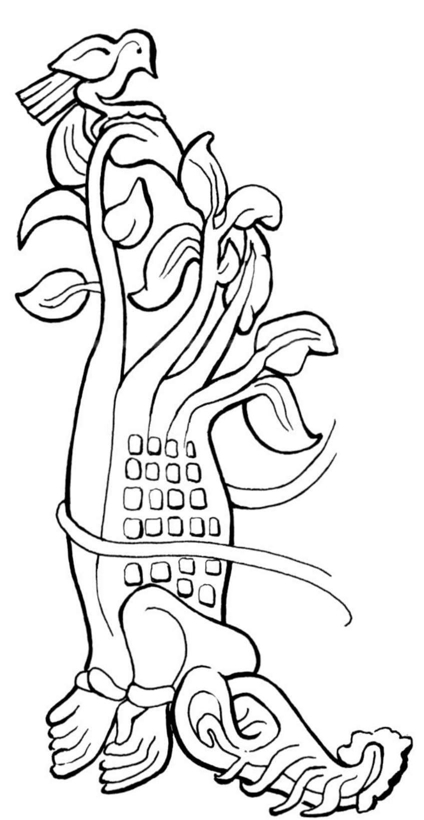

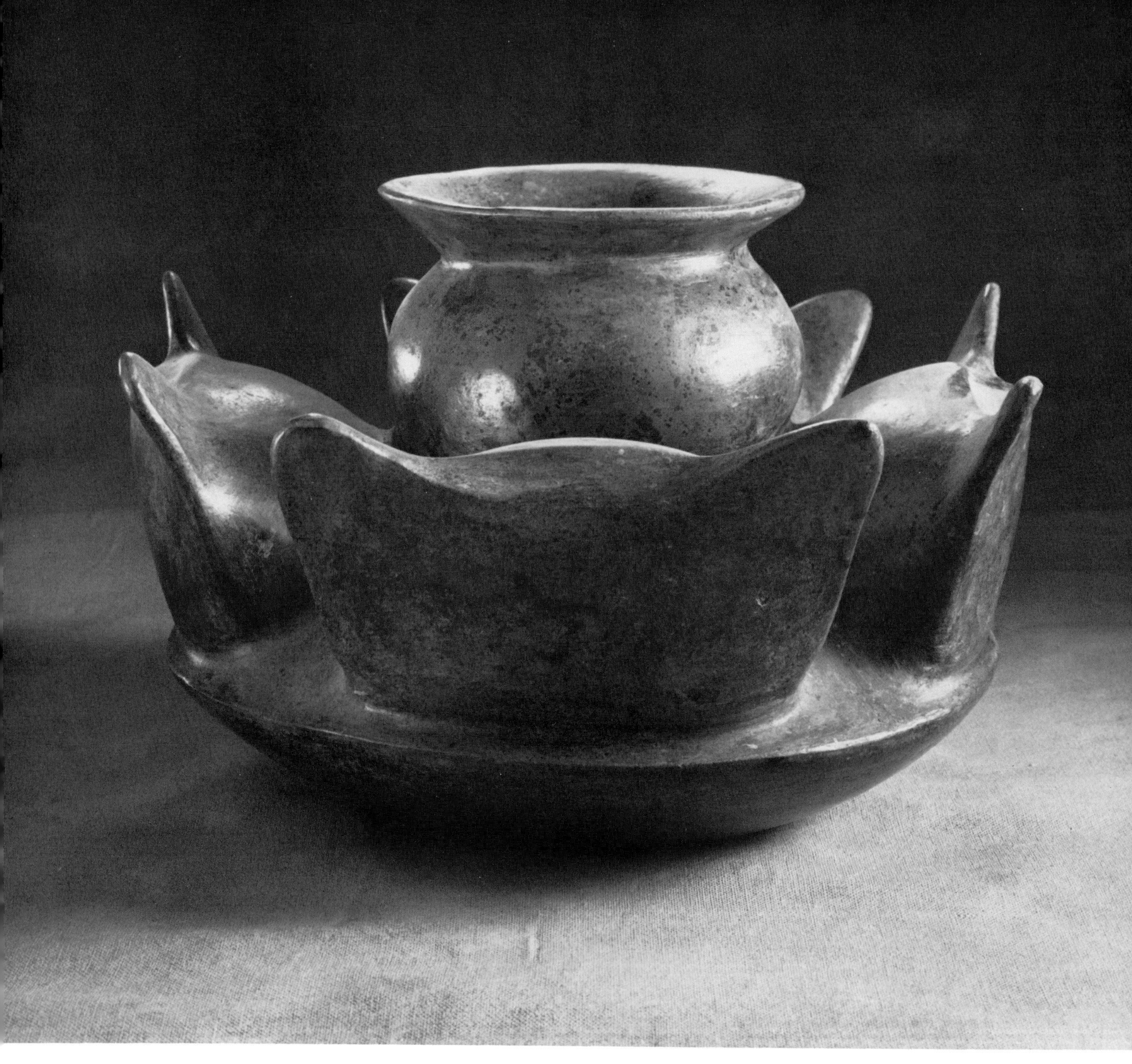

COLIMA "SEED POD" VESSEL 21
Colima, West Mexico
Late Preclassic–Early Classic: 200 B.C.–A.D. 300
Terracotta with burnished red-orange slip. H: 7⅛″
Lent by Constance McCormick Fearing,
Photo courtesy the Los Angeles County Museum of Art
L.83.11.951

CACTUS VESSEL 22
Colima, West Mexico
Late Preclassic–Early Classic: 200 B.C.–A.D. 300
Monochrome red. H: 7¾″
The Natural History Museum of Los Angeles County
F.A.774.71–8

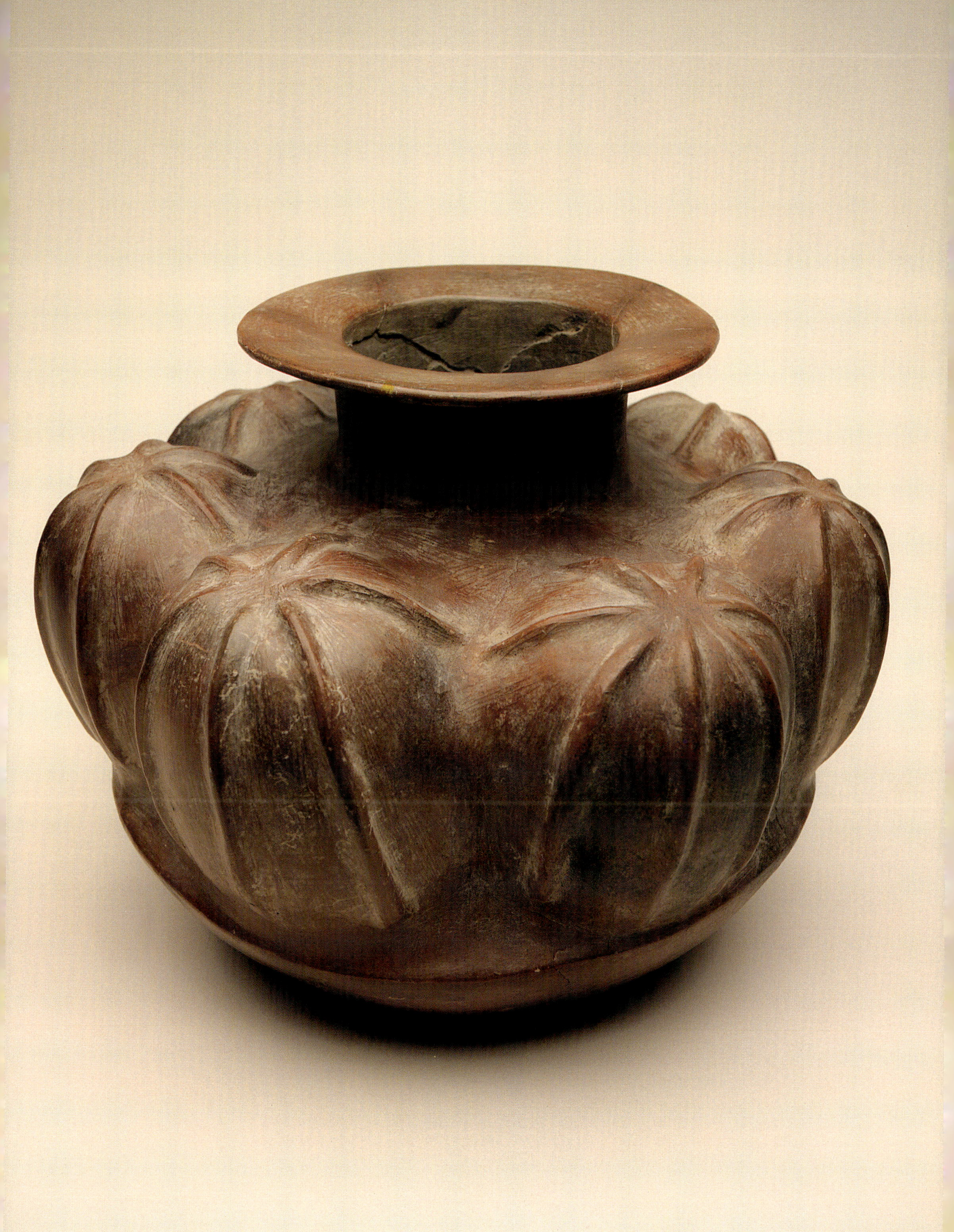

FROG — RANA   *tamazolin; bab*
TOAD — SAPO   *cueyatl; uo, wo*

An enormous toadlike monster, named Tlaltecuhtli (Earth-Lord) by the Aztecs, also served as a metaphor for the earth. Like the crocodile and turtle, this hideous toad floated on a primeval sea. Its gaping, toothy jaws swallowed the sun at dusk and consumed the sacrificial blood and hearts of victims offered the earth.

Both toads and frogs are associated with rain and the sustenance it assures. Throughout Mesoamerica, their croaking heralds rainstorms. Among the Maya, frogs are known as the rain god Chac's musicians. The choral announcements of frogs who "sing merrily" are noted by the Cora of Western Mexico; the Tarahumare speak of the toad as a "powerful rainmaker" (Lumholtz 1902, I:323,514). Precolumbian antecedents exist in the illustrations of the Maya *Codex Madrid* where frogs appear in rain imagery (folio 17c) or surround the rain deity Chac to the four directions (folio 31a).

Their connotations of fertility are strengthened by the fact that frogs and toads may lay thousands of eggs and assume a squatting position similar to that of a woman in childbirth. It is no wonder that parallels between these reptiles and maize or corn are many. In the Aztec celebration that heralded the rainy season, the offerings to Cinteotl (Corncob Lord) included a stiffly baked frog painted blue, the color of water and preciousness. For many Maya peoples today, small black frogs called *uo* are thought to have stomachs filled with new corn. The Giver of Maize for the Tzotzil lives in a mountain home and has his cave guarded by a frog.

Toads took on additional importance as they seemed to reenact the transforming changes achieved by shamans through mind-altering drugs (Furst 1981). The very life cycle of the toad, from egg to fish to a four-legged, carnivorous animal, is a dramatic metamorphosis. Toads also shed and consume their own skin several times a year.

The largest toad in Mesoamerica, Bufo marinus, is equipped with venomous skin glands. These glands secrete 27 different poisons including a hallucinogen (Fig. 28). There is evidence that the Olmecs farmed this hallucinogenic substance. At the Olmec site of San Lorenzo, archaeologists have uncovered large quantities of toad remains. Olmec objects, such as a green jade pallete in the form of a toad may have held hallucinogenic snuff (Joralemon 1988:48). The knowledge of the toad's powerful secretions persisted. According to a seventeenth-century traveler, Thomas Gage, Highland Maya peoples added toad to their potent brew of tobacco to increase its intoxicating properties (in Thompson 1970:120). Given the pharmacological importance of the toad, the Olmec were jaguar may actually be a toad with jaguar traits.[12]

CYLINDER VESSEL WITH PLANT DESIGN 23
Maya. Petén, Guatemala
Late Classic: A.D. 600–900
Polychrome terracotta. H: 4¾"
Stendahl Galleries

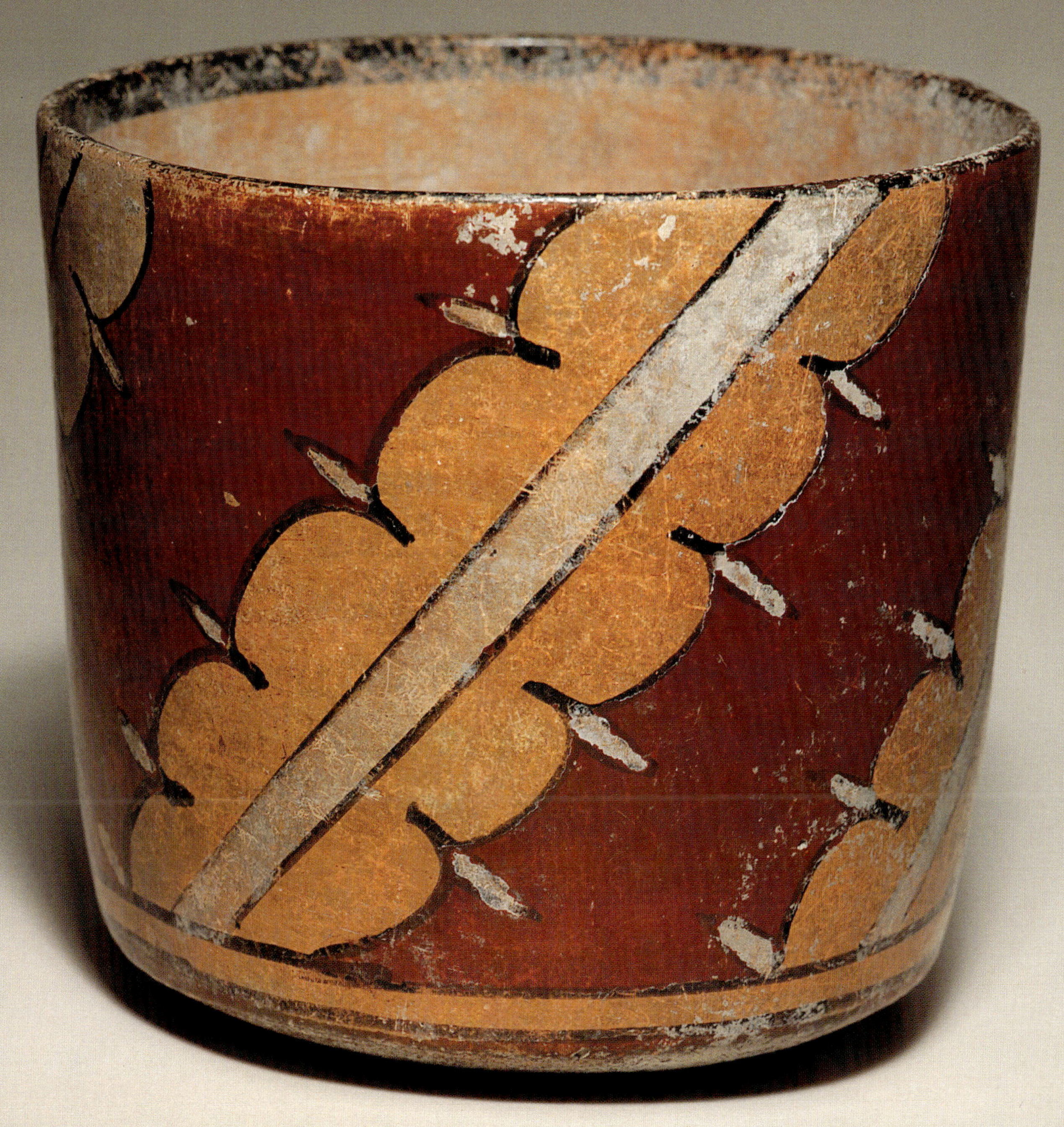

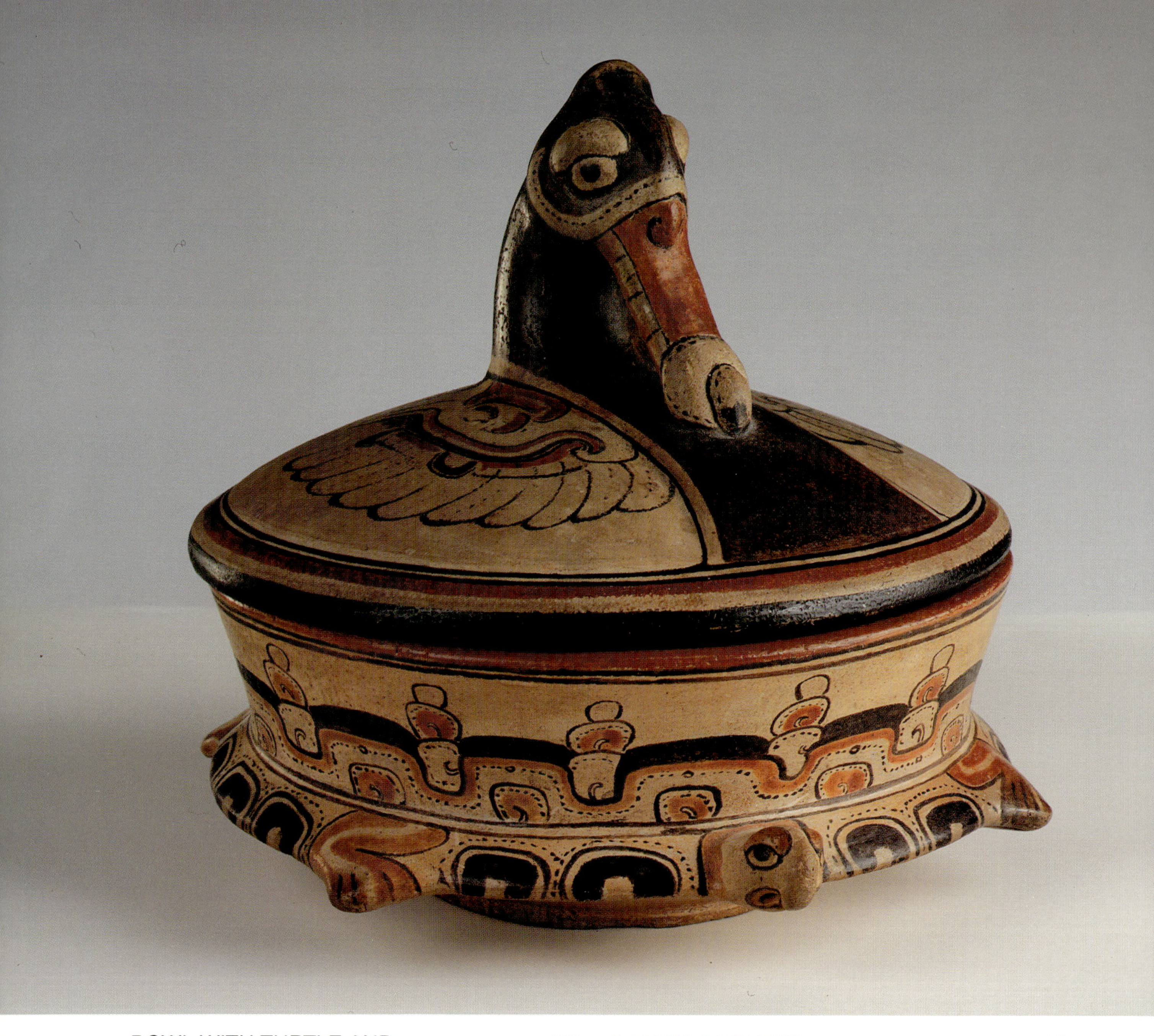

BOWL WITH TURTLE AND WATERBIRD LID   25

Maya. Tikal (Mundo Perdido), Petén, Guatemala
Early Classic: A.D. 350–400
Polychrome clay; buff with black, red, orange paint
H: 6⅔″
Museo Nacional de Arqueología y Etnología, Guatemala
11416 a & b

The projecting flange of this clay vessel depicts the turtle's carapace or upper shell. The turtle's head, feet, and tail are subtly modeled. It is swimming in water, symbolized by undulating dots, scalloped bands, and water stacks (illus. 9). On the lid is the head of a long beaked waterbird, serpent-wings painted on either side. The turtle was a metaphor for the earth's surface, whereas the waterbird could penetrate the various cosmic spheres in flying skyward and diving beneath the water. (In Clancy et al. 1985: Fig.45)

PAIR OF TURKEYS   24

Nayarit, West Mexico
Protoclassic: c. A.D. 0–500
Clay with cream slip. H: 6″, 6¼″
The Land Collection ML 509
(In Dwyer & Dwyer 1975: fig. 30)

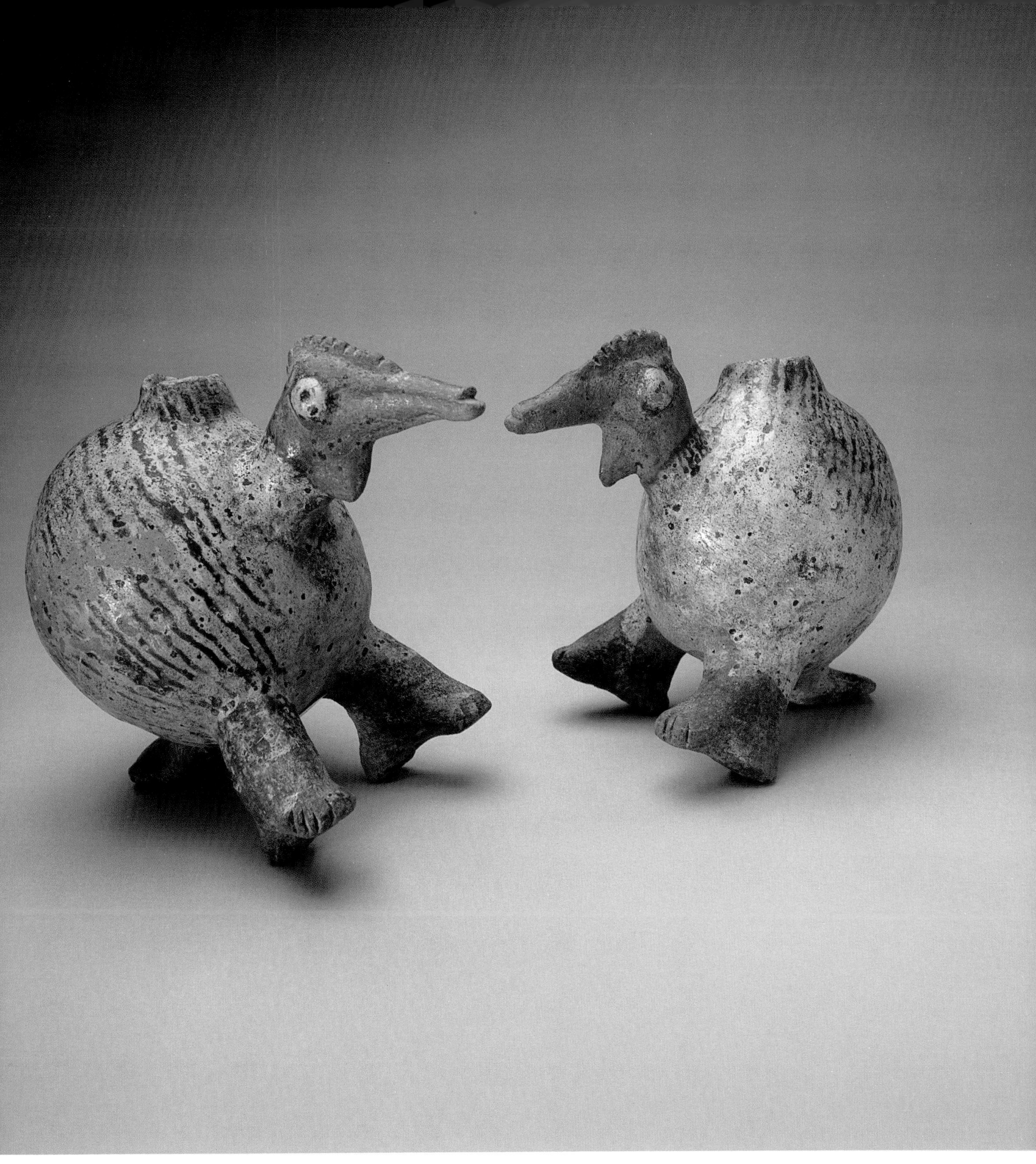

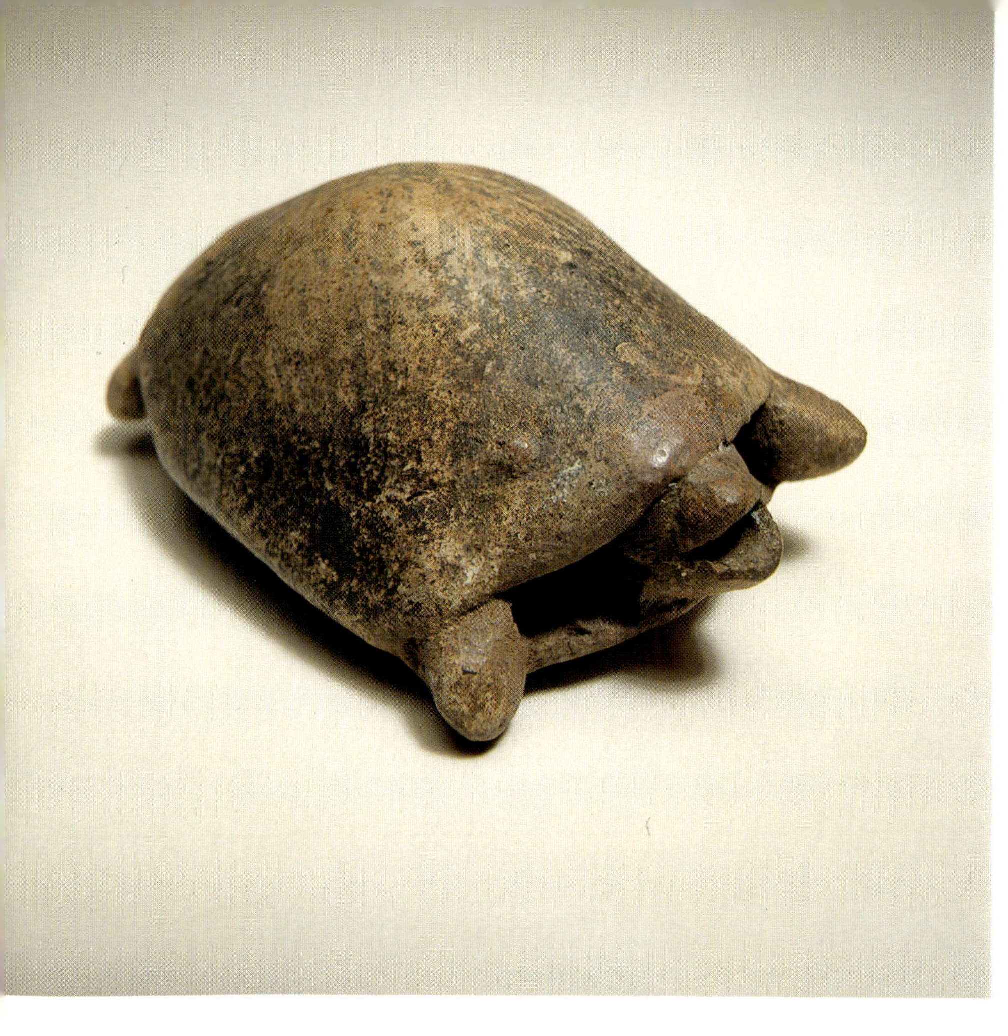

TURTLE WHISTLE 27
Colima, West Mexico
Late Preclassic–Early Classic: c. 200 B.C.–A.D. 300
Buff clay. L: 3″
Martha Roth

BOWL WITH FOUR TURTLES 26
Huastec. Veracruz, Gulf Coast, Mexico
Postclassic: c. A.D. 1200–1400
Buff clay with black and red on cream. H: 6½″
The Bowers Museum, Santa Ana
EL 79.9.8

This Huastec clay bowl features four turtles, whose heads and front feet project out laterally in four cardinal directions.

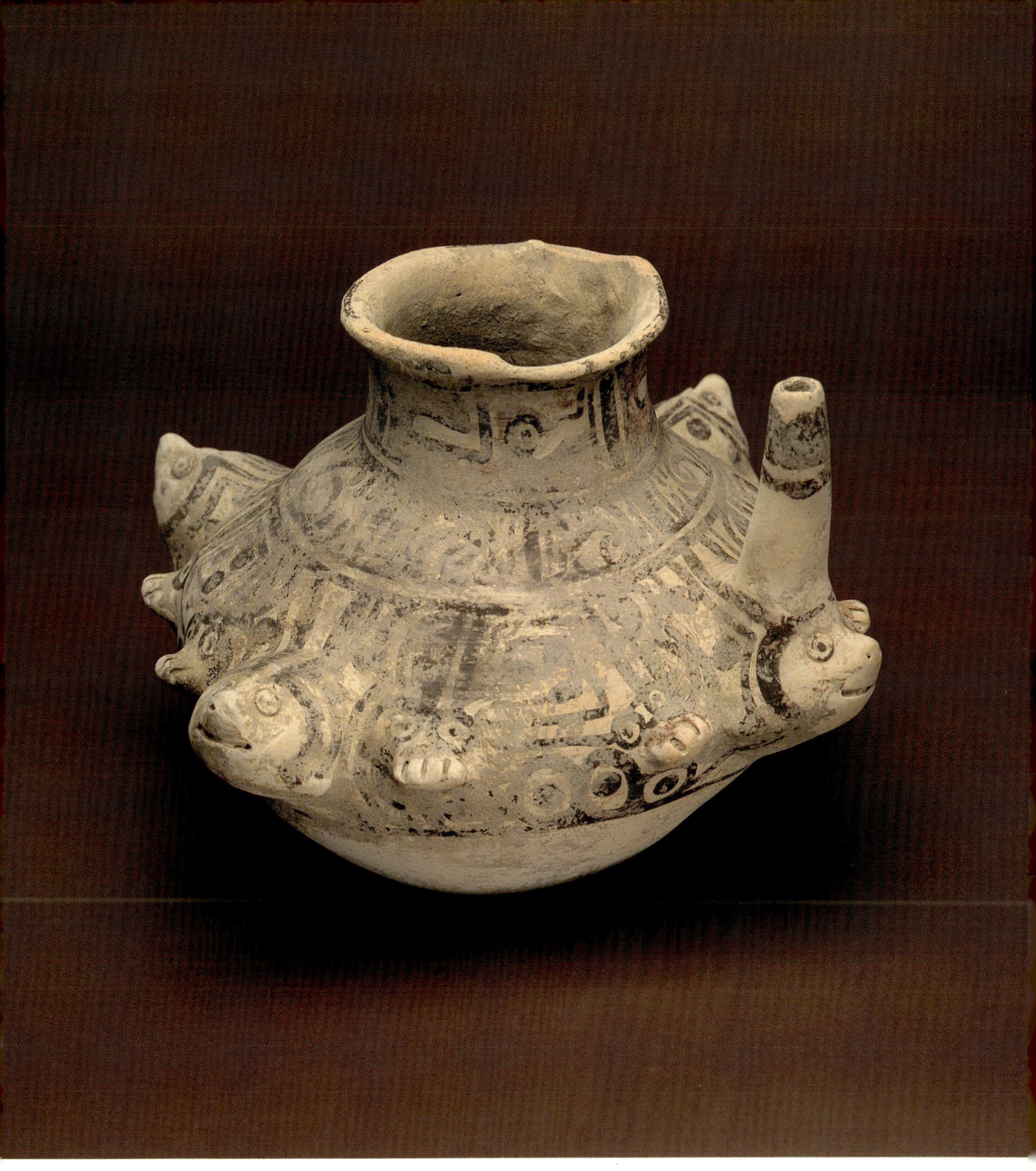

MONKEY — MONO, CHANGO
ozomatli; matz, batz, chouen

We are forever fascinated by the monkey, an animal in whose antics we see a parody of ourselves. In the monkey's face, upright posture, manual dexterity, and gregarious nature, we find a mirror of human behavior. As both mimic and contradiction of human culture, the monkey in the Precolumbian world mediated between man and the supernatural. The monkey was just as appropriately located in the sky because of its solar symbolism as it belonged to the netherworld of dead ancestors, but as patron of the arts, monkeys are discussed here in man's own earthly realm.

Monkeys were associated with an older mythic period of chaos that preceded this era of real men and cultural restraints (Benson 1989). Among the Postclassic Maya, there was an early time when the wooden people were destroyed by floods and became monkeys. These human mannikins first used tools and monkeys were a by-product of their designs; thus it is, explains the Popol Vuh, that monkeys today look like people.

The creative act itself was attributed to monkeys. In the Popol Vuh, the older half brothers of the hero twins are named One Monkey (Hun batz) and One Artisan (Hun chouen), sometimes translated as One Howler monkey and One Spider monkey. The half brothers are described as men of genius, as flautists, singers, writers, carvers. Once defeated by the hero twins and converted into monkeys, they became the deified patrons of the arts, "and writers and carvers prayed to them" (Tedlock 1985:124).

That monkeys were perceived as the originators of the performing and visual arts may be linked to their entertaining traits. The long-limbed Spider monkey (Ateles geoffroyi) displays the acrobatic grace of a dancer. The Howler monkey (Alouatta palliata) has a specialized larynx that can produce a wide range of hoots and cries, some of which boom across a distance of three forest kilometers. If the Howler is a master of sound, the dexterity of the rarer Capuchin (Cebus capuchin) recalls those skills needed for the crafts of sculpture and painting (Benson n.d.). Painted ceramics preserve many images of the monkey as the Classic Maya patron of the arts. Monkey scribes appear holding pens or brushes as well as paint pots. They were associated with the priestly and ruling class who claimed the privilege of learning hieroglyphics.

The monkey scribes' most common attribute was a large deer's ear. Like the deer, the monkey also had solar attributes. Monkey replaces deer in some Mayan myths where the sun and moon goddess mate. The day *ahau* glyph as young sun, may be represented by a monkey head; conversely, a personified form of the *kin* (sun) glyph is a monkey-man god.

In Aztec cosmogony, monkeys were also ancestral to mankind. The second Aztec era ended violently, with the survivors transformed into monkeys. This epoch was ruled by Quetzalcoatl under the sign for wind, Ehecatl. Monkeys shared Quetzalcoatl's insignia, such as his curved earrings and spiral shell pectoral. In the broadest sense, wind implied breath or life. Quetzalcoatl was a creator god who shed his own blood on bones gathered in the underworld to create the human race. Thus, monkeys and Quetzalcoatl were linked as vitalizing, creative powers (Figs. 29, 30).

The sign One Monkey in the Aztec calendar was presided over by Xochipilli (Flower Prince), a sun-related deity of flowers, games, and the arts. So it was that monkeys became associated not only with song and dance, but also with the excesses of pleasure, intoxication, gluttony, and lascivious behavior. Monkeys were clearly perceived as charmed creatures. In demand as imports, monkeys were kept as pets by the Aztec upper class to bring good luck.[13]

The animal's depiction in art with cacao pods may be explained by the more sensuous, even hedonistic side of monkeys (Figs. 31, 32; illus. i). Like monkeys, cacao (Theobroma cacao) thrives in warmer climates. From the fruit or seed pod of the cacao tree, Mesoamericans made a prized chocolate drink restricted to the wealthy upper class among the highland Aztec. The reputation of chocolate was not only as a beverage, but as a stimulant and aphrodisiac.

COATIMUNDI — PISOTE, TEJON *pitzotl; chi'ik, tziz*

The coati (Nasua narica) is a lowlands raccoon-like animal. Its prominently pointed nose and long tail held aloft, make it an easily recognizable mammal both in art and nature. The coati probes holes with its snout and digs in the ground; it satisfies a voracious appetite with insects, fruit, and even lizards. Effigy sculptures of the animals fre-

---

TOAD VESSEL   28

Maya. Mexico or Guatemala
Postclassic: A.D. 1000–1500
Plumbate Ware. H: 4⅞″
Lent by Constance McCormick Fearing,
Photo courtesy the Los Angeles County Museum of Art.
L.83.11.1025

Sculptures of the largest toad in Mesoamerica, Bufo marinus, often show the venomous skin glands conspicuously located behind and above the eyes. These parotid glands secrete a potent hallucinogen that was used ritually through the colonial period.

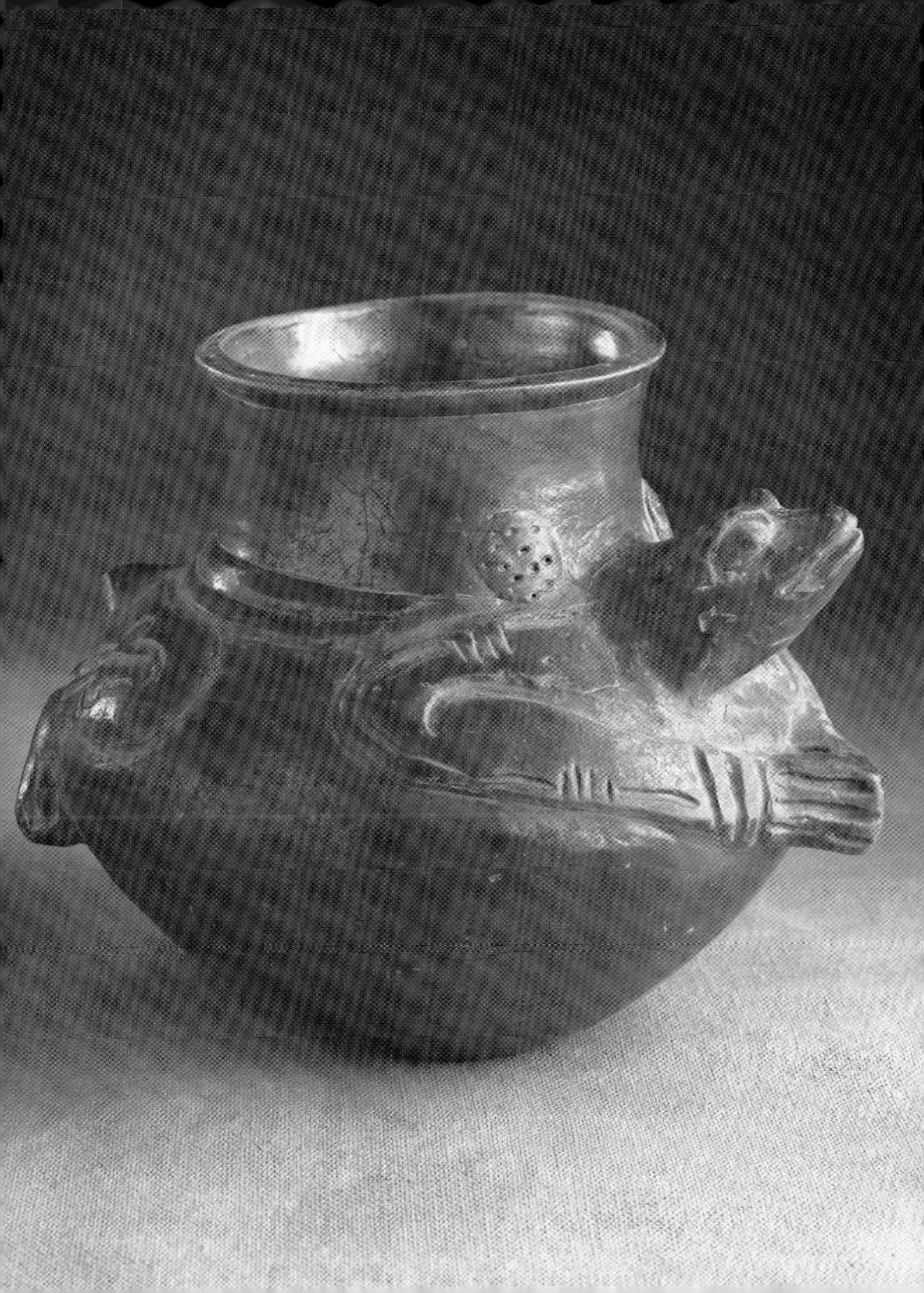

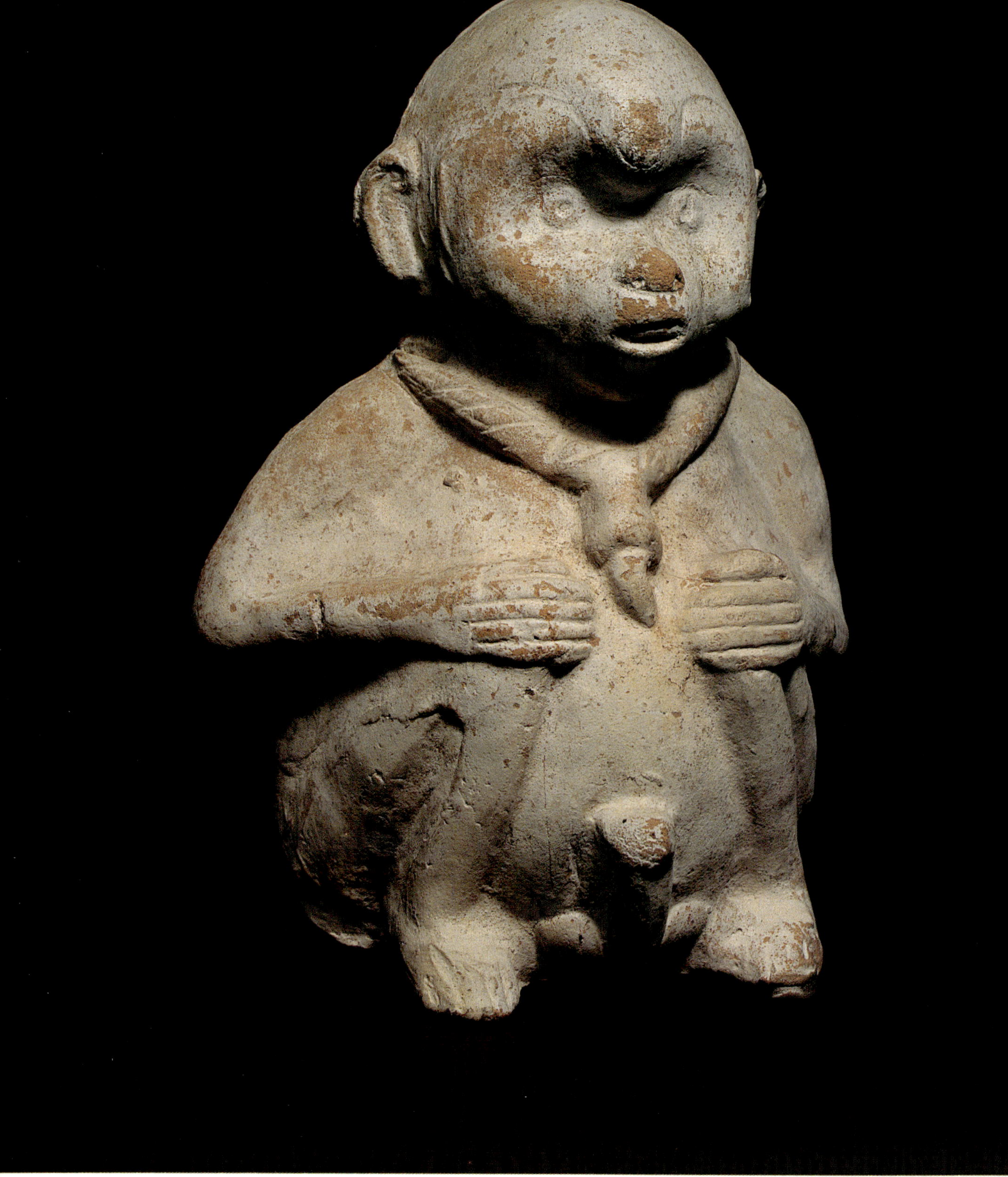

MONKEY EFFIGY 29
Maya, Jaina-style. Campeche, Mexico
Late Classic: A.D. 550–800
Buff clay with white paint. H: 8″
Stendahl Galleries

MONKEY EFFIGY VESSEL WITH LID 30
Maya. Kaminaljuyu, Guatemala
Late Preclassic: c. 500 B.C.–A.D. 100
Grey-brown terracotta with fire clouds. H: 12¾″
Collection of The Denver Art Museum, gift of M. Larry
Ottis, M.D. 1987.511, a & b

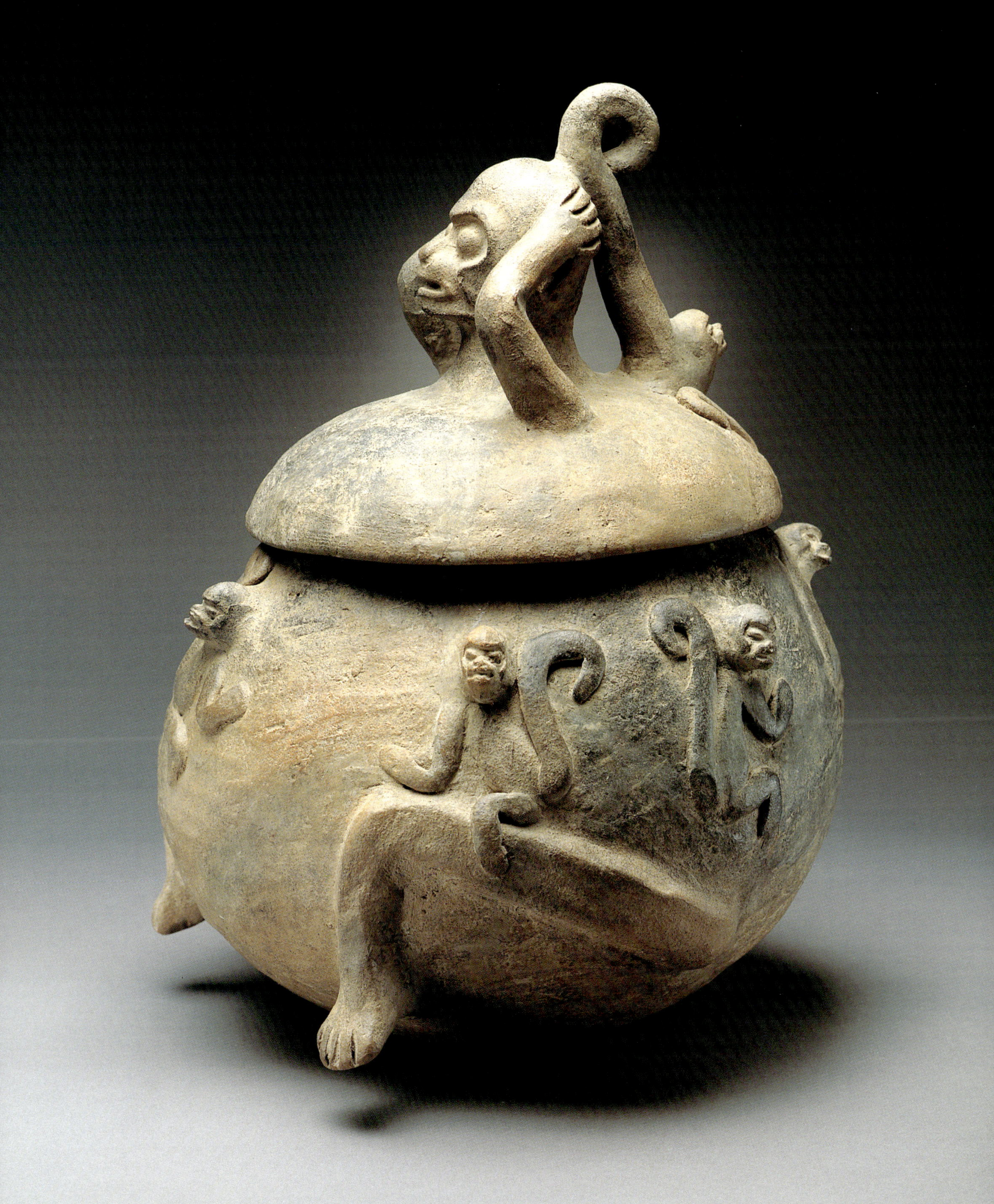

quently show it eating, often holding fruit or corn (Fig. 34).
Easily tamed, the coati's funny antics make it a desirable
pet. The coati's foraging and playfulness help explain its
close associations with agriculture and ritual clowning.
The coati is one of several furry animals prominent in contemporary agricultural rites. These are celebrated by
traditional cultures from the Southwest United States
down through Middle America and appear to have great
antiquity (Redfield 1936). Associated pole-climbing rituals
associated with these rituals include participants dressed
as sacred clowns who hold or wear furred animals, either
monkeys, squirrels, or most frequently, the coati. A coati
can also be displayed on top of the tree-pole itself,
described as a World Tree. It is likely the coati replaces
the maize-dough image that surmounted the pole
erected by the Aztec for their celebration of Xocotl huetzi
described above. In this century, coati were suspended
with fruit and vegetables in front of Guatemalan churches
to insure a good crop. The Quiche creator god,
Cucumatz (Plumed serpent), is also called "coati," in
addition to other agricultural names.

ARMADILLO — ARMADILLO   *ayotochtli; huech*

Our English name for the armadillo is taken from the
Spanish for "little armed one" or *armado,* because of its
tank-like protection. Although encased in horny plates,
the flesh of the armadillo is very edible; hence, its Nahuatl
name, *ayotochtli,* that combines turtle *(ayotl)* with rabbit
(tochtli).

Although all twenty species of armadillos are found in the
New World, nine-banded armadillos (Dasypus novem-
cinctus) predominate in Mesoamerica. And, in fact, the
nine jointed plates around the animal's midriff are in
themselves significant for their reference to the sacred
number nine and the nine underworld levels. In the Maya
region today it is still one's good fortune to meet up with
a nine-banded armadillo (Fig. 35). The cosmic number
four is likewise attached to armadillos, since they consistently bear quadruplets of the same sex! Like the coati,
the armadillo was, and continues to be, a sign of earthly
fertility.[14]

GOPHER — TUZA   *tozan; ba*

The gopher shares the armadillo's powerful claws with
which it not only forages in the ground for roots and
insects but digs an underground burrow (page 9). With
edible flesh, like the armadillo's, and the ability to store
food in its cheek pockets, the gopher was also a creature
associated with the earth's bounty. In contemporary
Maya creation myths, the gopher replaces the monkey
as the second race of beings and four pairs are sometimes located at the four world directions. On the other
hand, as an essentially subterranean animal, the gopher
was a sign of the buried dead for both the Aztecs and
the Lancandon Maya.[15]

THE PRECOLUMBIAN RUBBER BALLGAME

Plant and animal imagery associated with the rubber
ballgame reinforced the Mesoamerican duality of life and
death, sky and underworld, wet and dry seasons. In its
ancient form, the rubber ball "game" was a ritualized
contest with political and religious implications. Demand-
ing physical prowess as well as courage, the sport was
dangerous, for a direct hit was potentially fatal. Further-
more, it was sometimes played for the highest stakes, with
the outcome jeopardizing the very life of the contestants.
By the time of the Spanish conquest, the ballgame was
also held for purposes of recreation and gambling.

Vestiges of this widespread ballgame can still be found
in the state of Sinaloa in Northwest Mexico (Leyenaar
1978). Called Ulama, the name of this game retains the
Nahuatl word for rubber, *olli* or *ulli*. *Olli* was, in turn, based
on the Aztec word for "movement" or *ollin*. It was the very
springy quality of the new rubber material that so
fascinated the Europeans. Unlike the leather balls from
the Old World, rubber was remarkably elastic.

The Rubber tree (Castilla elastica) from which latex is
extracted flourishes in hot, wet lowlands. The ballgame
first developed on the Gulf Coast of Mexico, an area rich
in rubber-bearing trees. Its cultic importance reached an
apogee from A.D. 400–700 when the ballgame appears
to have accompanied the trade routes for cacao or chocolate, rubber, and cotton.

The ballgame replayed a widespread myth in which the
sun god descended to the underworld to be reborn with
the maize god. Keeping the ball aloft was a metaphor for
the sun's journey, both above and below the earth. In the
underworld, the land of the dead, paradoxically, lay the
genesis for life; thus, the transformation that occurred in
the underworld sustained both agricultural and com-
munal life.

BOWL WITH MONKEYS AND FRUIT PODS 31
Maya. Petén, Guatemala
Late Classic: c. A.D. 600–800
Polychrome clay. H: 5⅞"
The Land Collection
(In Dwyer & Dwyer 1975: Fig. 86)

Two long-limbed spider monkeys chase each other
around this vessel, holding fruit pods that may be from
the cacao (chocolate) tree.

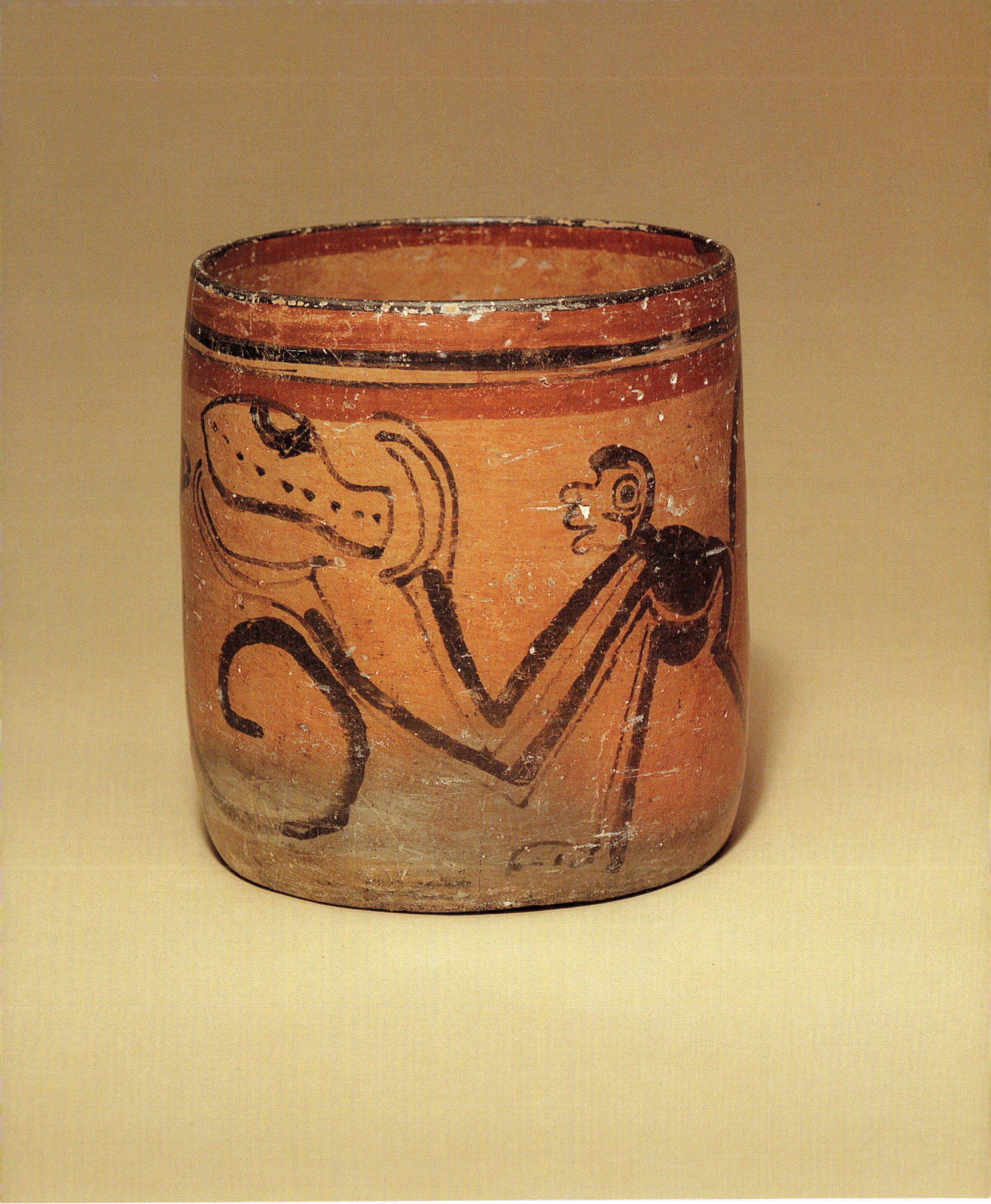

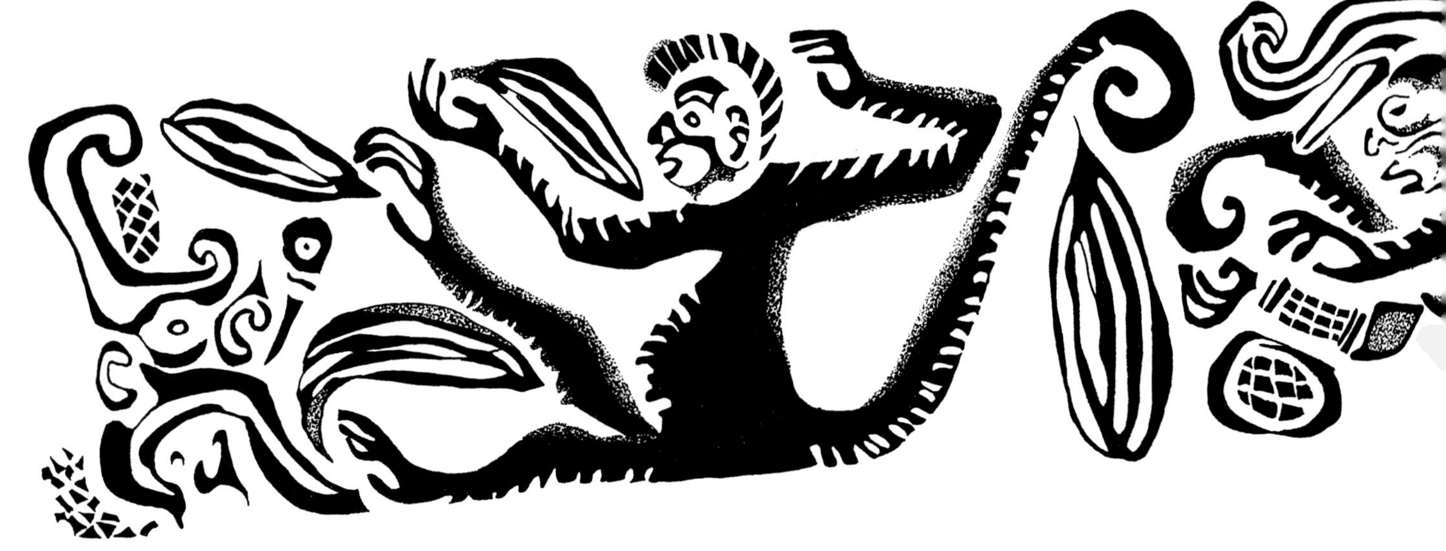

MONKEYS WITH CACAO PODS ON VESSEL

32

Maya, Peten or Guatemalan Highlands
Late Classic: c. A.D. 600–900
Incised and carved blackware. H: 5″
Collection of The Denver Art Museum, gift of M. Larry Ottis, M.D. 1989.214
Photograph courtesy of Denver Art Museum

Illustration drawn and inked by Kelly Poche

Both a spider and a howler monkey as well as a squirrel are carved in relief on the sides of this cylinder vase (illus. i). A human figure is intertwined with one of the monkeys. All animals are shown holding or playing with oblong, striped fruit that resembles cacao pods. The chocolate derived from cacao was made into a special beverage known to be a stimulant and drunk especially during ritual occasions. It was perhaps a mutual reputation for festivity and sexuality, that brought cacao and monkeys together.

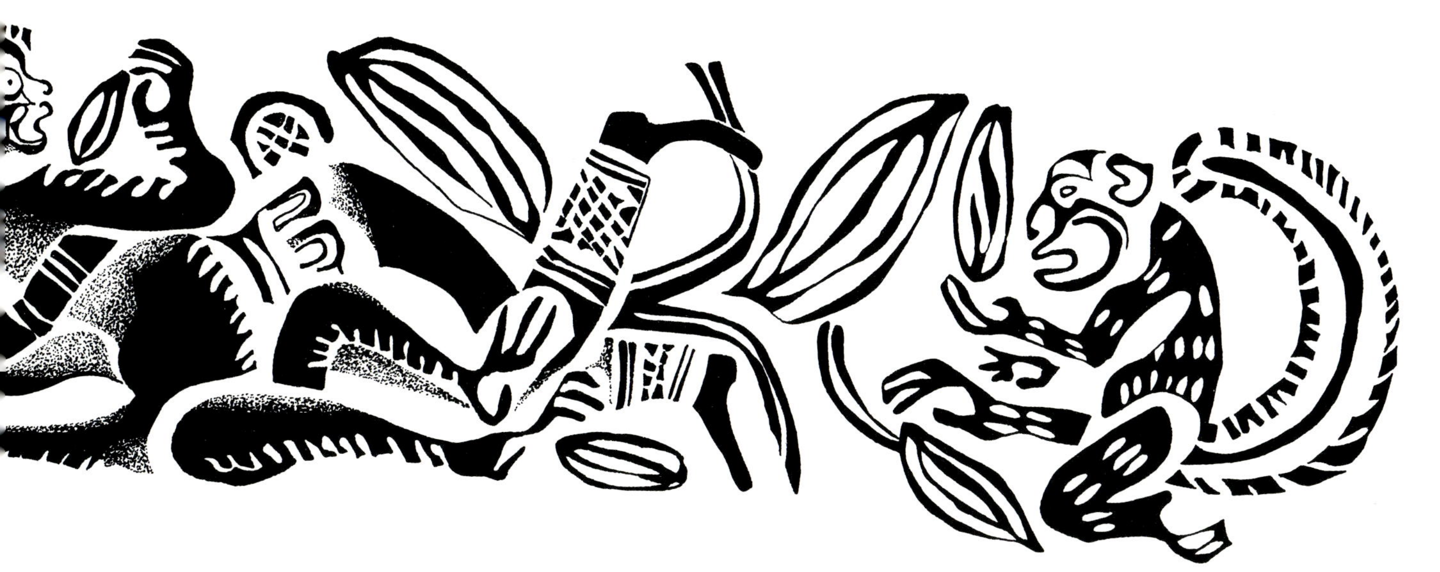

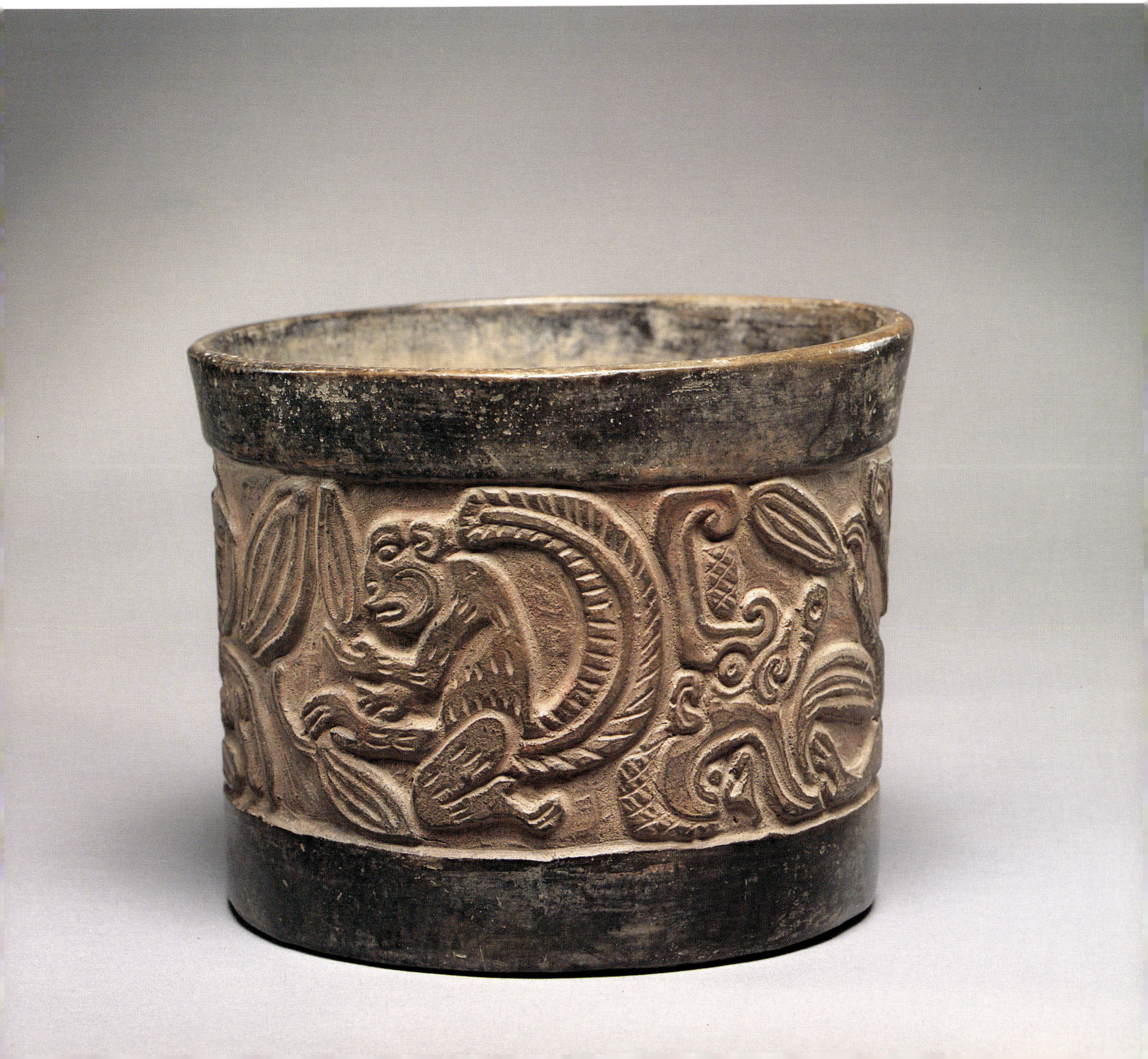

An integral part of the game included blood sacrifice, principally in the form of decapitation. The team, or its captain, who lost the game, lost their lives as well. In scenes of decapitated ballplayers, streams of blood are converted into serpents or flowering vines. The use of snake imagery as a blood substitute reinforced the concept of sacrifice as a corollary to life. At the level of a state cult, ballgame rituals reinforced royal lineages and predetermined the outcome of military campaigns.

Stone masonry ball courts have been excavated throughout Mesoamerica and are generally shaped like a capital "I." A narrow alley is bounded by vertical or sloping side walls. The size of the courts varies from 20 meters in length all the way to the super bowl size of 150 meters. The orientation of these courts to the cardinal directions points to the game's cosmic significance.

Rules for playing and keeping score varied from region to region. The most important aim was to keep the ball in motion. Although it could be bounced off of the sloping aprons, penalties arose when it was allowed to touch the ground. In most games, once the ball was in play, the players were required to hit it only with their hips or buttocks and heads. The solid rubber ball today weighs some five pounds and poses great physical danger to the players who must wear protective belts made of deerskin.

Precolumbian players likewise had to guard against internal injury by wearing one or more of the following gear: belts, kilts, knee pads, gloves, arm or waist bands, headdresses and special helmet-like headgear, and a type of closed sandal. Intricately carved objects were replicas in stone of more functional equipment made from perishable wood, leather, or wicker. Stone yokes *(yugos)* are U-shaped or oval objects that imitated protective belts worn around a player's waist or torso.

Attached to the fronts of the yokes were two types of objects used to help deflect the ball (Fig. 37). *Hachas* ("axes") are thin stone heads, perhaps derived from axes or blades. The taller, palm-leaf shaped objects are called *palmas* or palmate stones. Scenes of ballplayers suggest that the heavy stone yokes, hachas, and palmas were only worn ceremonially, before or after the game itself. Hachas additionally may have been portable markers for the ballgame, team emblems, or trophies. Their discovery in burials indicates that these were objects of great value.

Yokes, hachas, and palmas were carved with human, plant, and animal images appropriate to the cosmological and fertility meanings of the ballgame. Yokes from the Gulf Coast most commonly have toad or jaguar features (Fig. 38). These two animals signified the earth monster or possibly the entry to the underworld (Leyenaar & Parsons 1988:47). Yokes from the Maya area also convey strong earth symbolism as they were most frequently decorated with serpent imagery. The very con-

figuration of the U-shaped yoke may have symbolized "serpent" (Elayne Marquis, personal communication).

In addition to human and skull imagery, the largest number of hachas on the Gulf Coast represent underworld species, such as jaguars (Frontispiece), serpents, and iguanas. In the Southern Mesoamerican area, the preponderant number depict jaguar, monkey, bat, and deer. Oddly, toads and serpents, the two most common animals used for the yokes, appear only rarely on hachas.

Stone palmas display the greatest diversity of images including many composite animals, such as deer-headed serpents. Palmas also frequently focus on themes of fertility, with images of vegetation, particularly a palm-like leaf and corn, blood streams or serpents, and sacrificial scenes. Although not yet fully understood, flora and fauna motifs on the yoke-palma-hacha complex refer to various aspects of the earth's fertility, the entry to the underworld, and the underworld itself.[16]

FIGURE IN MONKEY COSTUME WITH FLOWER HEADDRESS 33

Maya. Jaina, Campeche
Late Classic: A.D. 550–800
Buff clay with white and blue paint. H: 13″
Stendahl Galleries

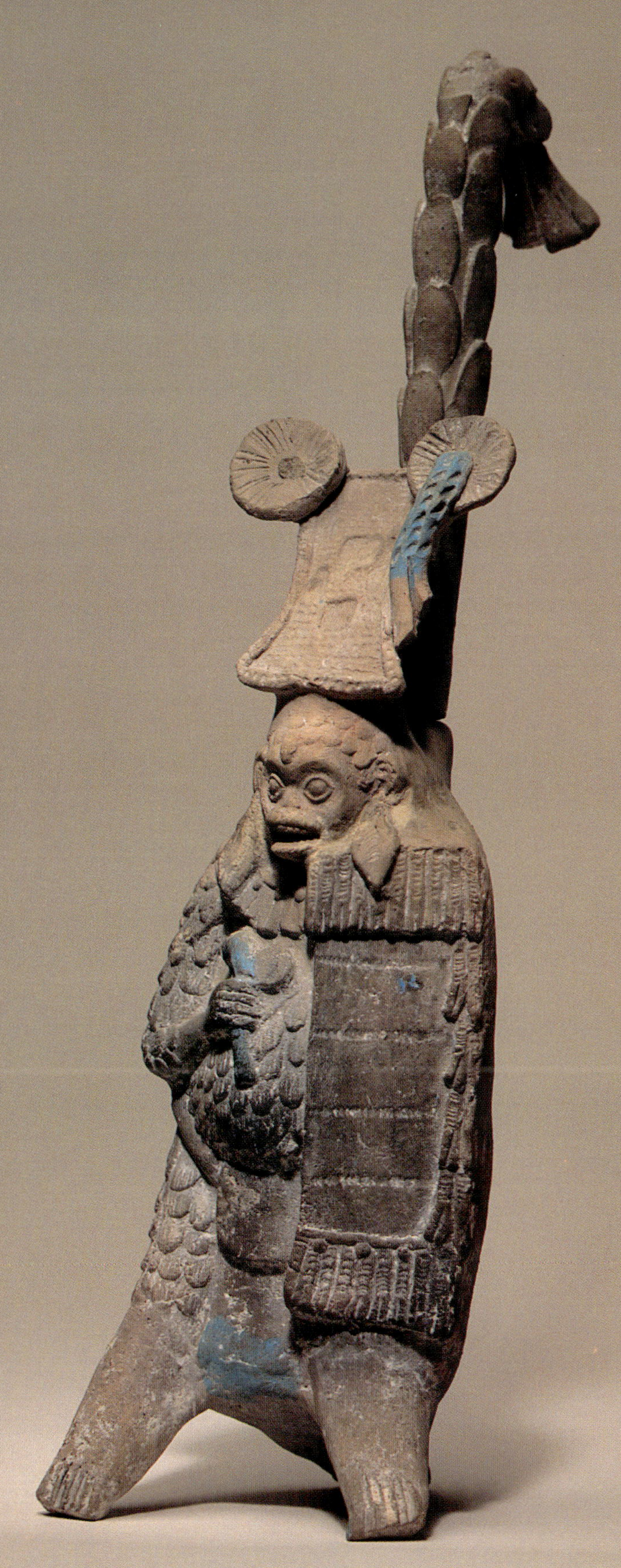

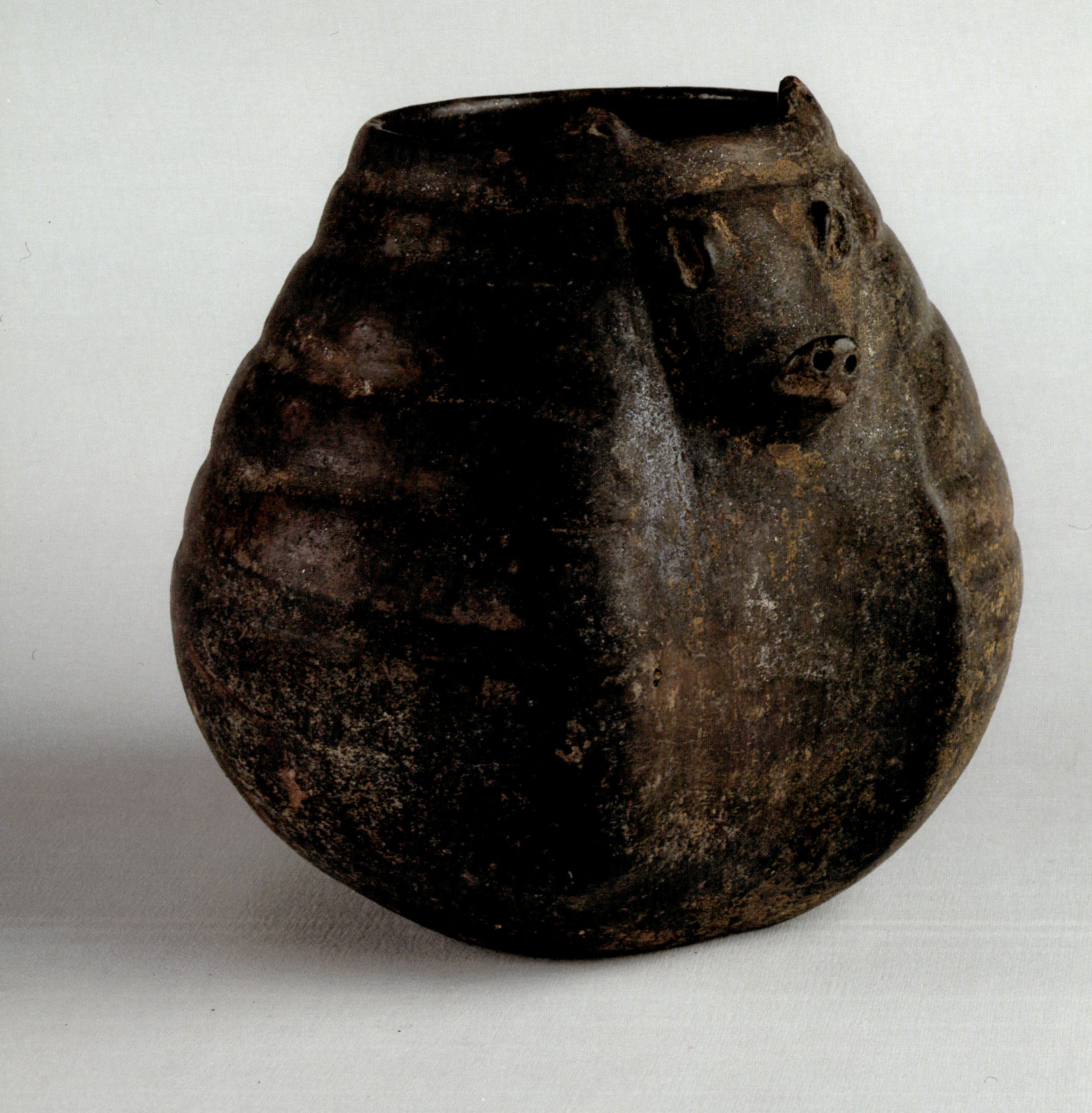

VESSEL IN FORM OF ARMADILLO 35		**COATIMUNDI EFFIGY** 34
Maya. Kaminaljuyu, Guatemala		Colima, West Mexico
Late Preclassic: c. A.D. 100–300		Late Preclassic–Early Classic: 200 B.C.–A.D. 300
Blackware. H: 5½″		Bichrome burnished clay. H: 7¼″
Museo Nacional de Arqueología y Etnología,		The Land Collection ML 909
Guatemala 3458		

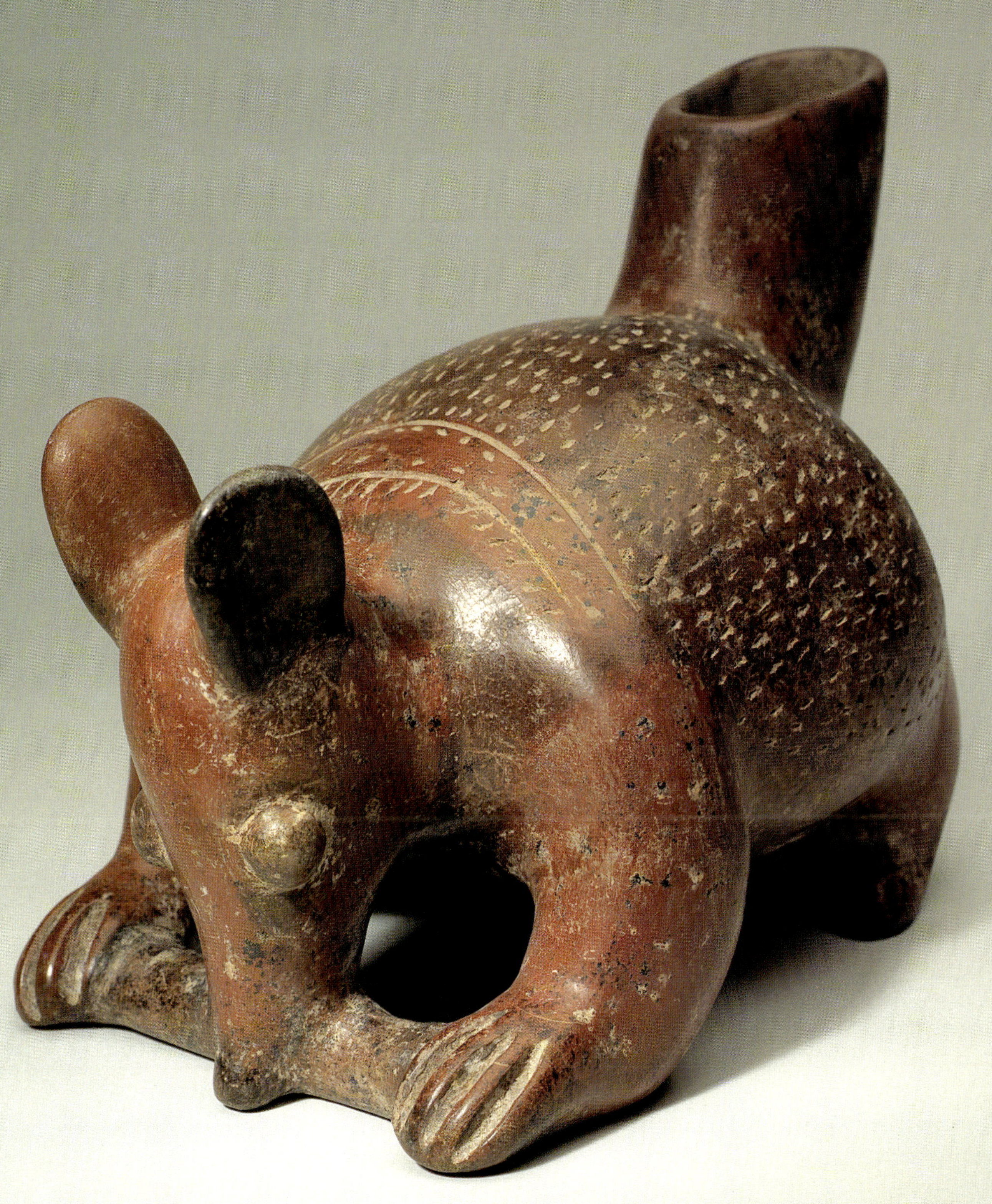

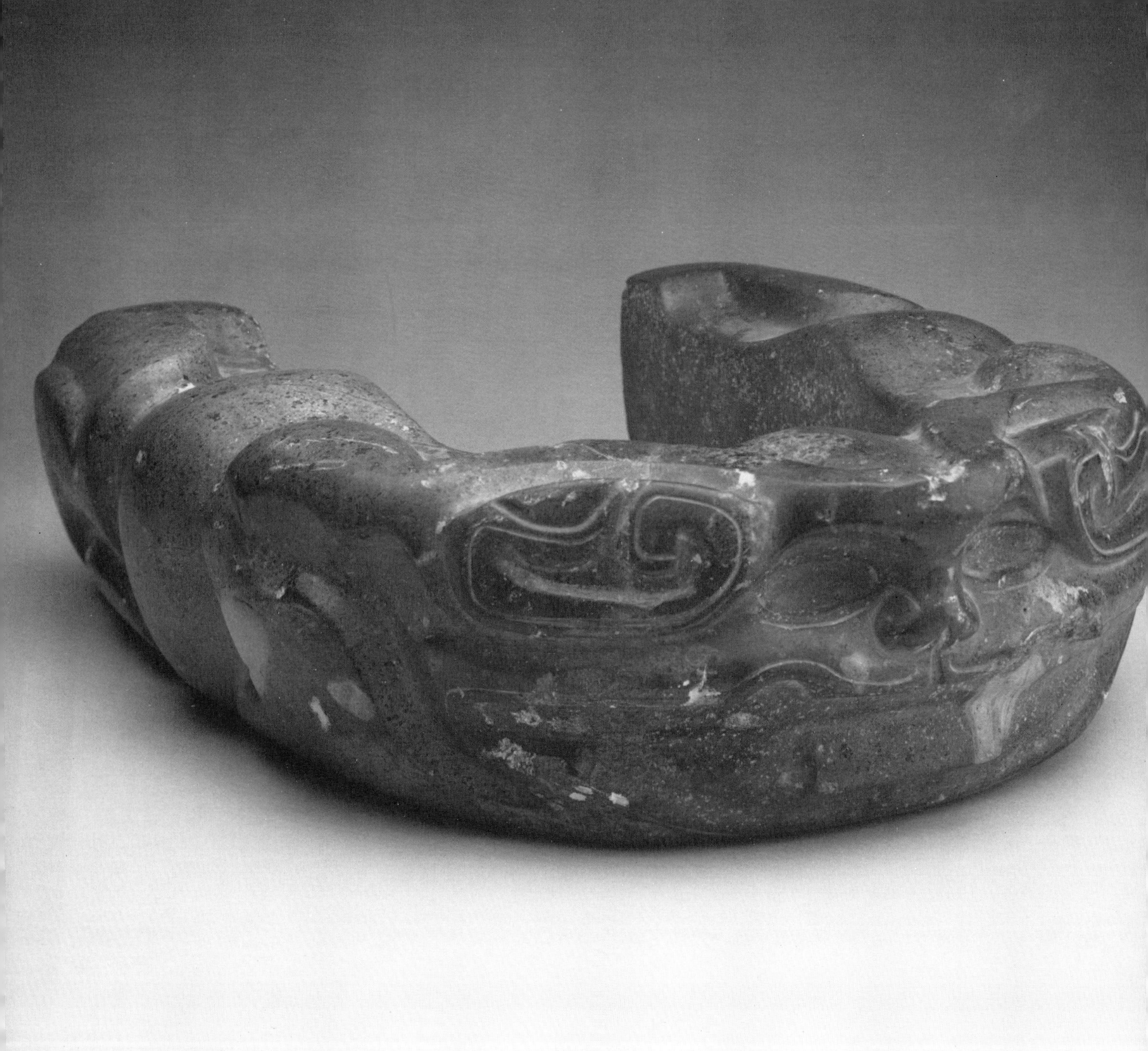

YOKE WITH MAN-JAGUAR DESIGN   38
Classic Veracruz. Veracruz, Gulf Coast, Mexico
Late Classic: c. A.D. 600–900
Grey-green diorite. L: 17″
Lent by Constance McCormick Fearing,
Photograph courtesy the Los Angeles County Museum
of Art L.83.11.875

As seen in figure 37, ballplayers wore U-shaped yokes
(yugos) of animal hide or wicker and padded with cotton.
Stone replicas of these yokes were ceremonial and
valued enough to have been buried with the ballplayers.
Yokes were carved with earth and underworld animals,
most commonly with the toad, jaguar, and serpent.

STANDING BALLPLAYER   37
Maya. Jaina, Campeche
Late Classic: A.D. 550–800
Buff clay with traces of white. H: 5⅞″
Stendahl Galleries

Since the ball in the ballgame was of heavy, solid rubber
and players primarily used hips and thighs to strike the
ball, they wore protective belts called yokes. This yoke is
decorated with a crocodile like head. The Jaina ballplayer
may have originally worn a deer headdress.

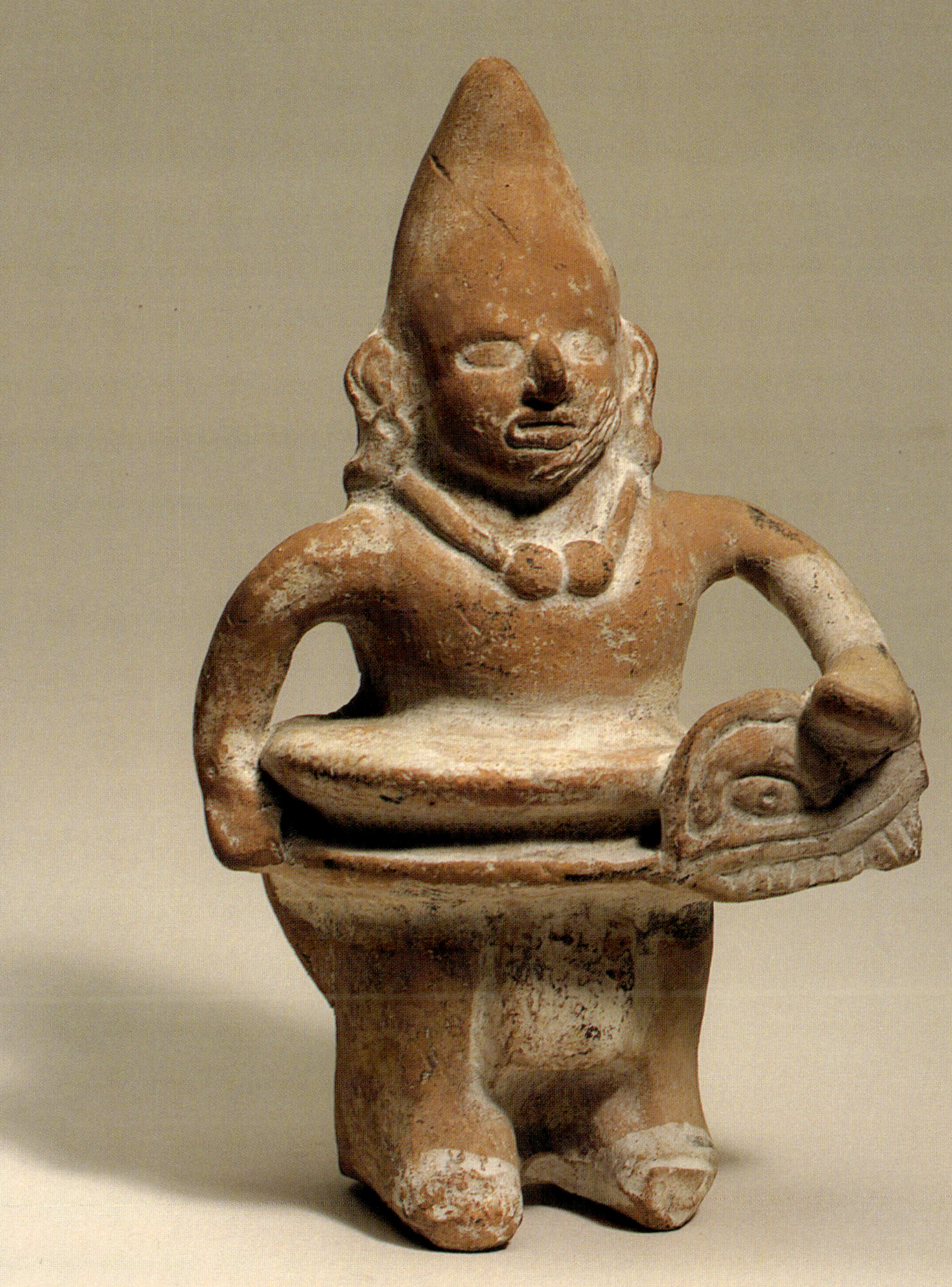

EARTH TO UNDERWORLD

DOG–PERRO   *Itzcuintli; pek*

In life and death, the dog was Precolumbian man's companion. Dogs were kept as pets, guardians of the home, and hunting aids. Of the four domesticated animals in ancient Mesoamerica, the dog was the most important. The others were the turkey (Fig. 24). Muscovy duck, and the bee.

The favorable tenth day sign was Dog in both the Mayan and Aztec calendars. Under the Aztec sign Dog, an individual was certain to be successful, well endowed with friends and material goods. Dogs were bred and sold commercially to be fattened and eaten. In the sixteenth century, Duran (1971:278–279) marveled at the hundreds of dogs of all sizes that were up for sale in the central Aztec marketplace.

Duran also lamented as idolatrous the use of dogs not only as meat for special banquets, but also as sacrificial animals. In many Precolumbian cultures dogs were sacrificed to honor various deities. Often the heart was extracted and blood smeared on the god-effigies; later the canine meat was cooked and eaten. Colonial reports indicate that dogs continued to be sacrificed long after human sacrifice was outlawed.

Curiously, it is unclear exactly which breeds existed in the New World. The Chihuahua is generally cited as the most common type of dog (Canis familiaris). More mysterious is the identity of the *tepescuintli* or hairless Mexican dog *(pelon mexicano)*. It has a higher body temperature than most dogs and provided a source of warmth. Was the hairless dog also the counterpart for the supernatural Xolo-itzcuintli, "Monster dog" or "Twin dog"? Was this the dog who acted as guide for the soul?

According to Nahua beliefs, the faithful dog accompanied his master into the underworld. Consequently, dogs were buried with the deceased, a cotton cord tied around its neck to lead the soul across the nine underworld rivers. Widespread among present-day Mayan cultures is a similar belief that a dog aids one in crossing a body of water before reaching the final underworld destination. The Lacandon Indians place palm figures as symbols of dogs at each corner of their burial mounds. Archeologists have uncovered dog bones within Classic Mayan tombs. When dogs were represented as emaciated, with prominent ribs and vertebral column, their skeletal appearance suggest their supernatural role as guardian for the dead (Fig. 41).

The dog's roles as guardian of the hearth and the dead were carried over to deities that had canine attributes. Dogs were closely associated with fire and particularly the fire of the domestic hearth. The goddess of the hearth fire, Chantico (In the house), was given the calendric name Nine Dog. In the Aztec creation myths, only dogs survived the destruction wrought by a rain of fire. A dog was also the animal form of Xolotl (Monster), twin of Quetzalcoatl. This canine deity was sometimes depicted with skull face and in the codices, with a death's head. Aztec sculptures of Xolotl included human ears with Quetzalcoatl's curved earrings and his own dog ears cut short into knobs (Fig. 42).

The largest single group of ceramic dogs come from the shaft tomb cultures of ancient West Mexico. Extraordinarily varied in body form and position, these dogs are represented alone, in couples, or grouped with human figures and other animals. Sizes range from the over-lifesized (Fig. 43) to miniature representations. Colima dogs in particular, are famous for their realism, vivacity, and smooth, rounded contours. Some appear to dance; others squirm on their backs or gnaw on a corncob (Figs. 40, 44, 45). These playful poses reveal the intimacy between dog and human.

Dogs were frequently included in sculptural "family portraits." In Preclassic examples from central Mexico, human figures (generally but not exclusively female) hold dogs, often in the same manner as a human baby (Fig. 48). Since dogs also provided nourishment, it is possible that in these works dogs connoted fertility. In other contexts, dogs take their place next to masters of some status (Fig. 47) or have been adopted as insignia of the Colima hunter-warrior (Fig. 46).

TWO PLAYFUL DOGS   **40**
Colima, West Mexico
Late Preclassic–Early Classic: 200 B.C.–A.D. 300
Burnished terracotta chrome red
H: 8¼"
The Land Collection ML 767
(In Nicholson & Cordy-Collins 1979: fig. 79)

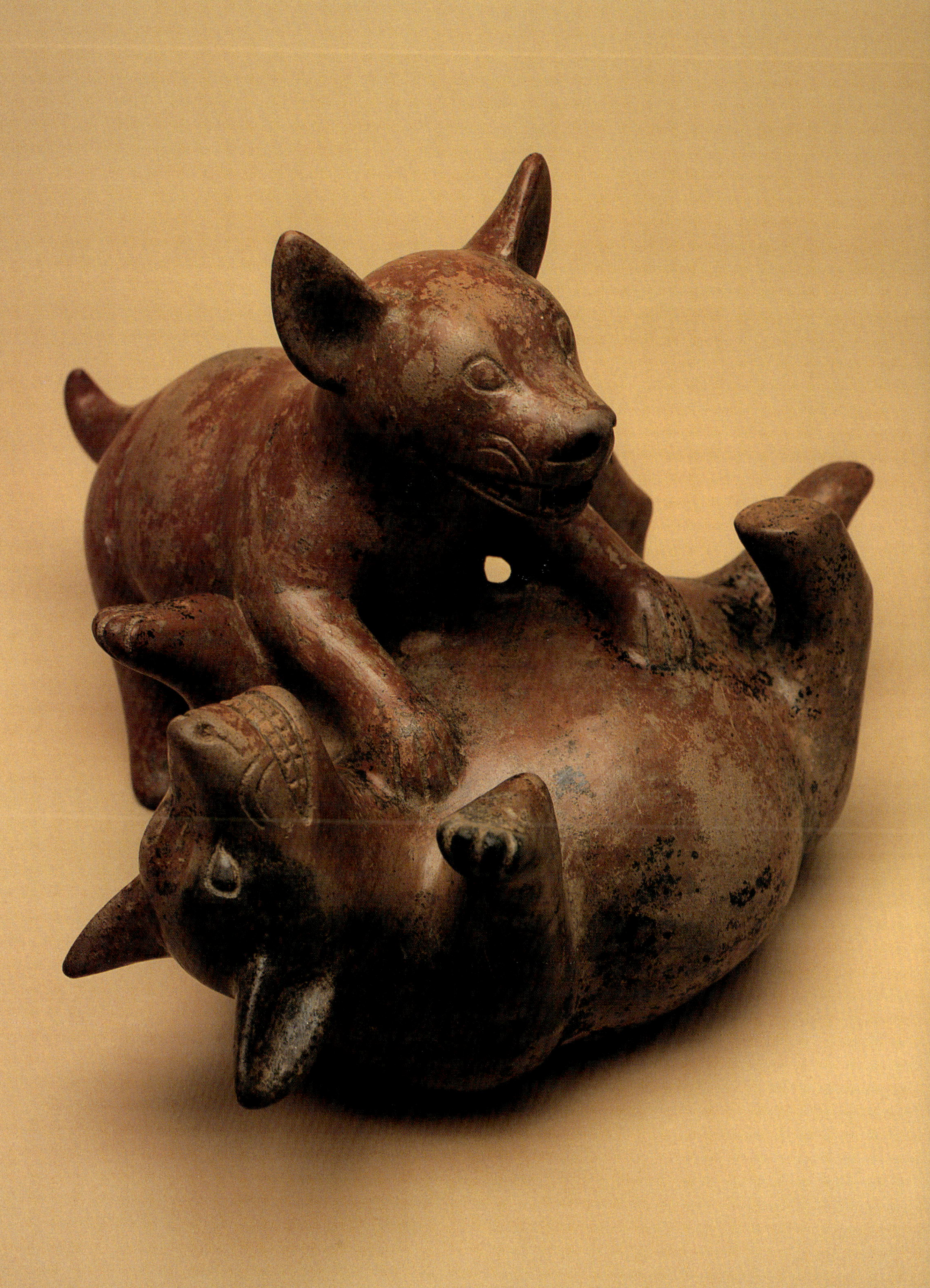

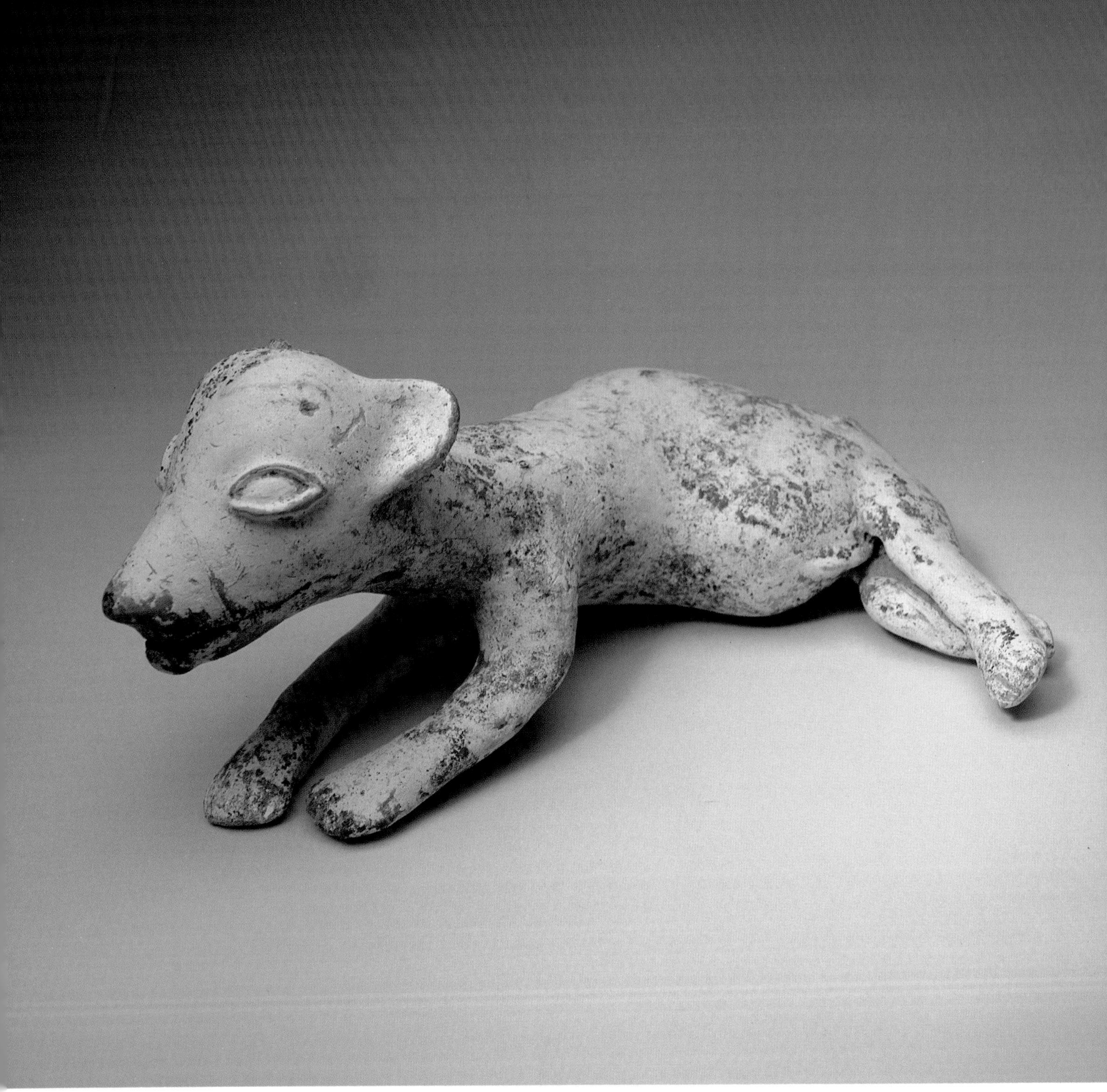

EMACIATED DOG EFFIGY 41

Jalisco, West Mexico
Protoclassic: 100 B.C.–500 A.D.
Buff clay with white slip. H: 3¾″
Mr. William Schneider

EARLESS DOG ON 42
INCENSE BURNER LID

Maya. Uaxactún (?), Guatemala lowlands
Mid Classic: c. A.D. 450
Blackware with pressed textile design. H: 5½″
Collection of The Denver Art Museum, gift of James
P. Economos 1972.201

A death associated Aztec canine monster deity named
Xolotl was characterized by torn dog ears. Xolotl, as
Venus the evening star, pushed the sun into the underworld on the Western horizon. This representation of a
black dog with amputated ears may be a Maya Xolotl.

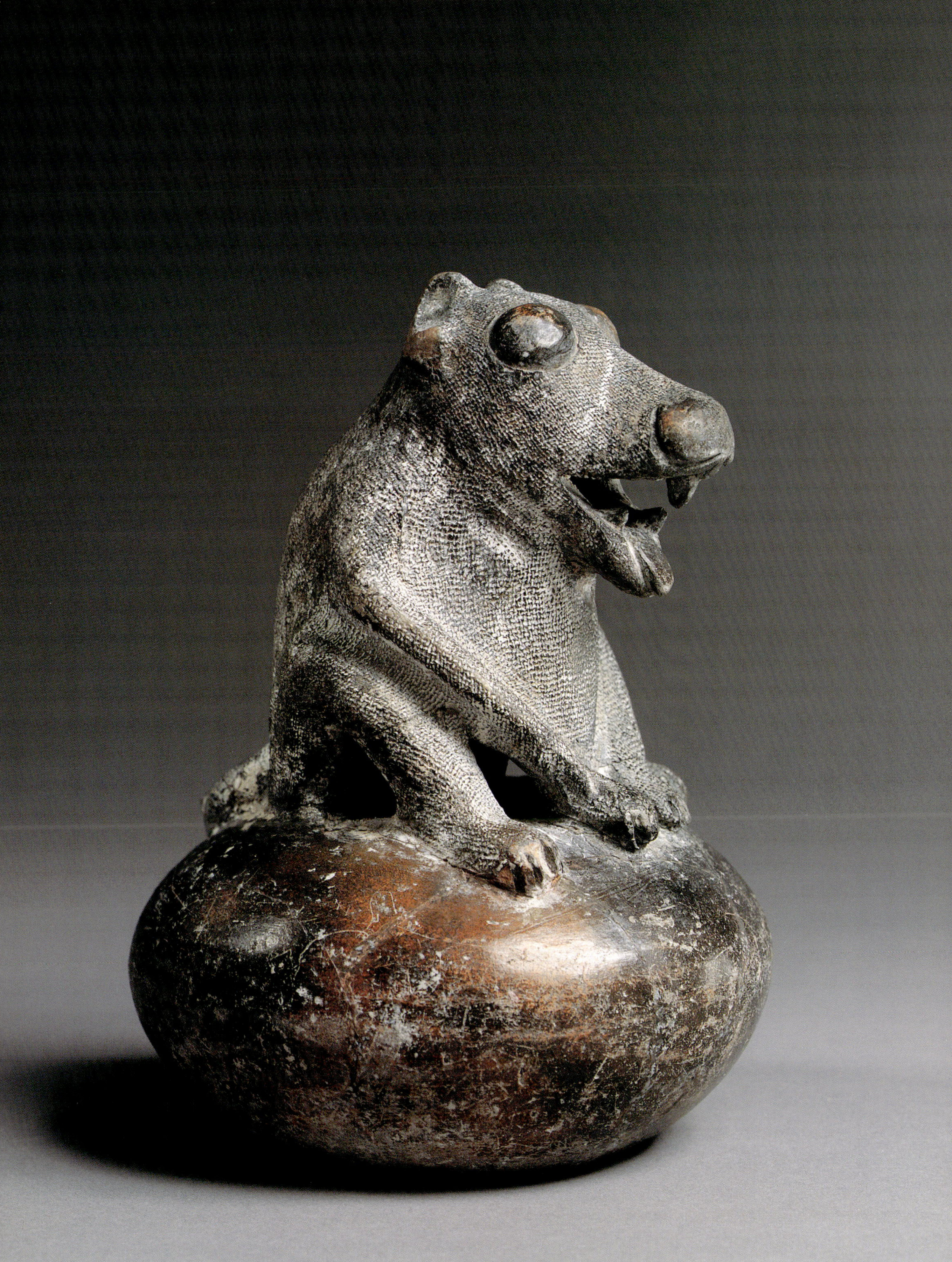

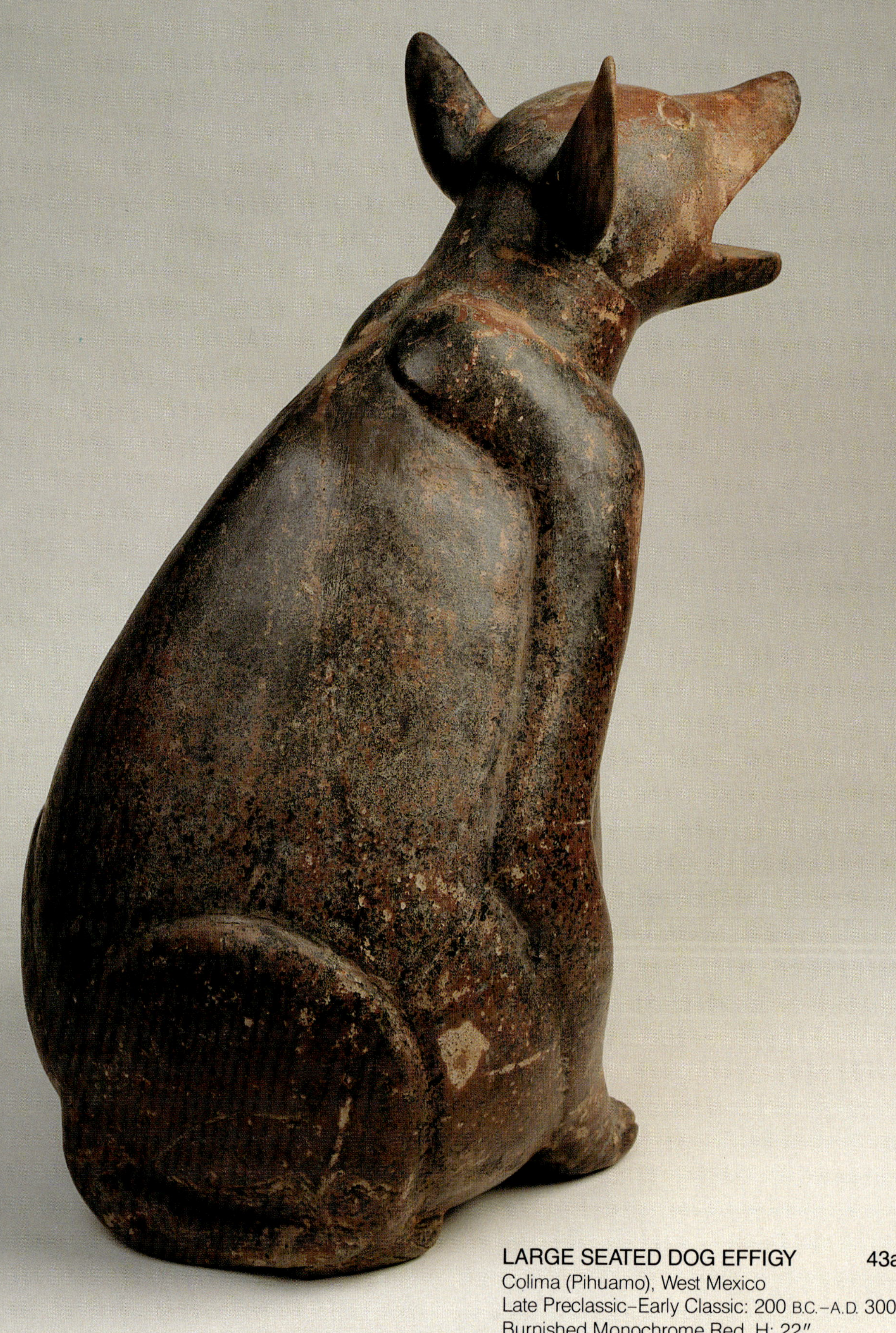

LARGE SEATED DOG EFFIGY 43a, 43b
Colima (Pihuamo), West Mexico
Late Preclassic–Early Classic: 200 B.C.–A.D. 300
Burnished Monochrome Red. H: 22″
Mrs. Jarvis Barlow

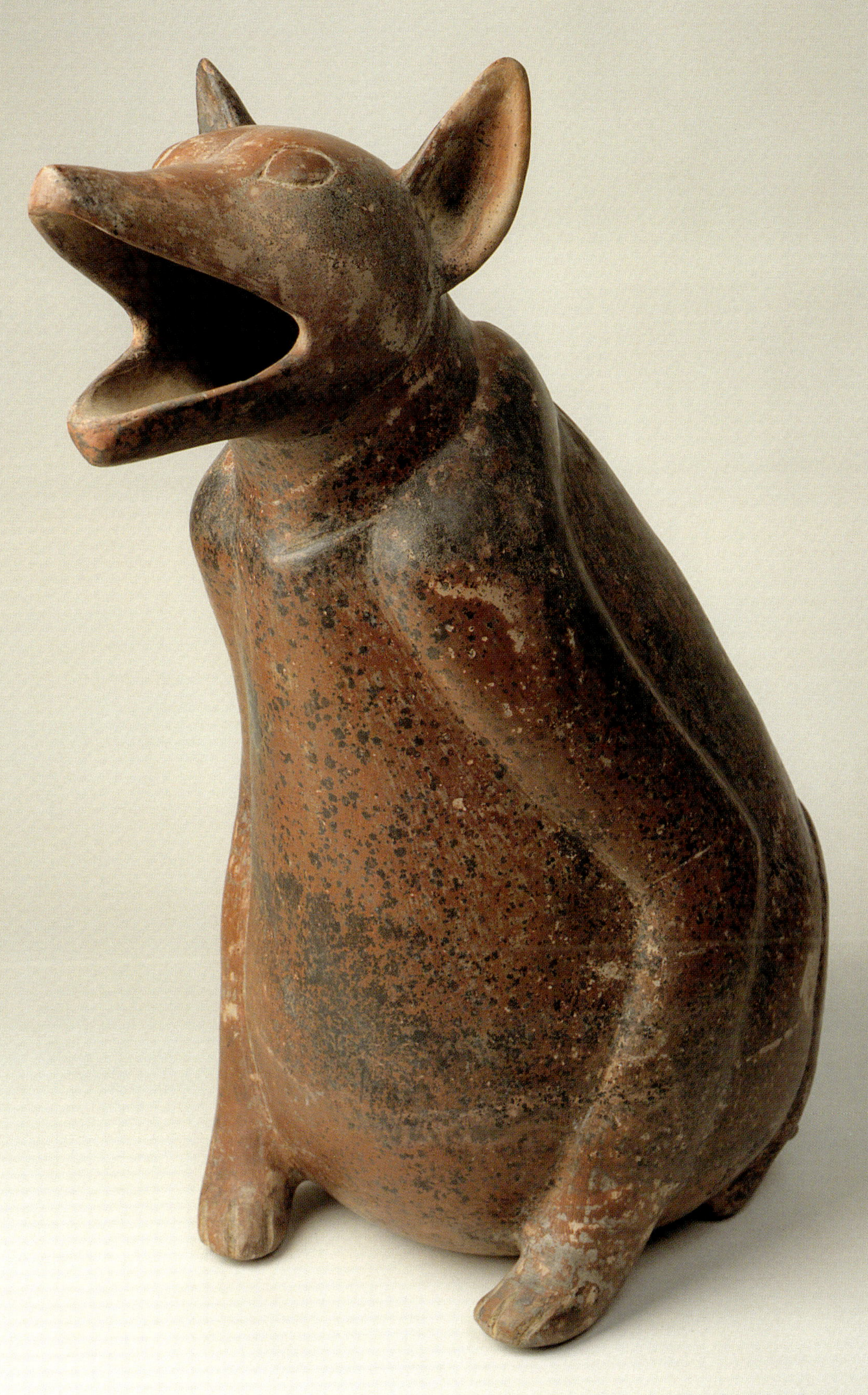

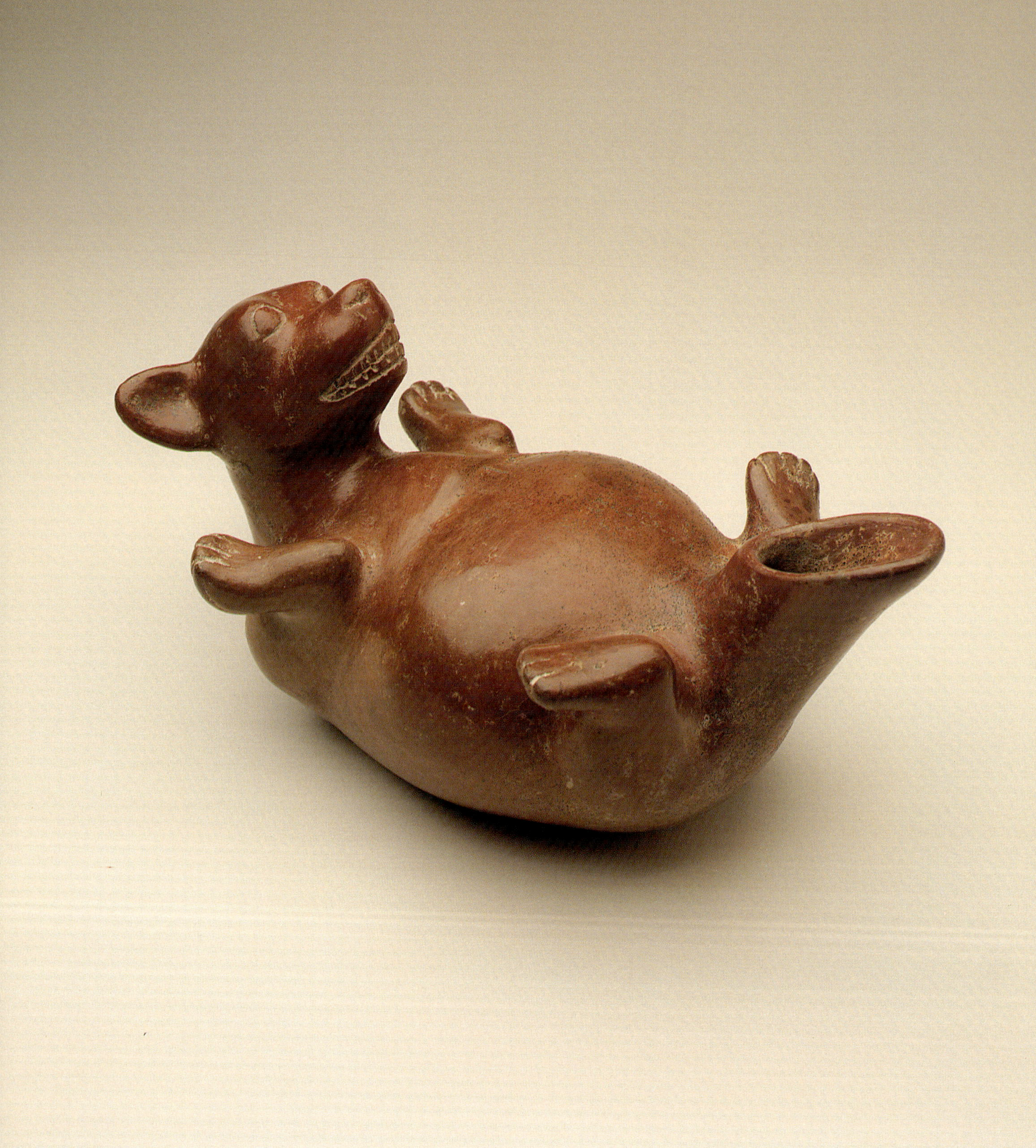

DOG LYING ON BACK 44
Colima, West Mexico
Burnished terracotta: bichrome tan and red slip painted
H: 6″
Mr. William Schneider

DOG WITH CORN COB 45
Colima, West Mexico
Earthenware, burnished red slip.
H: 6″
Private collection

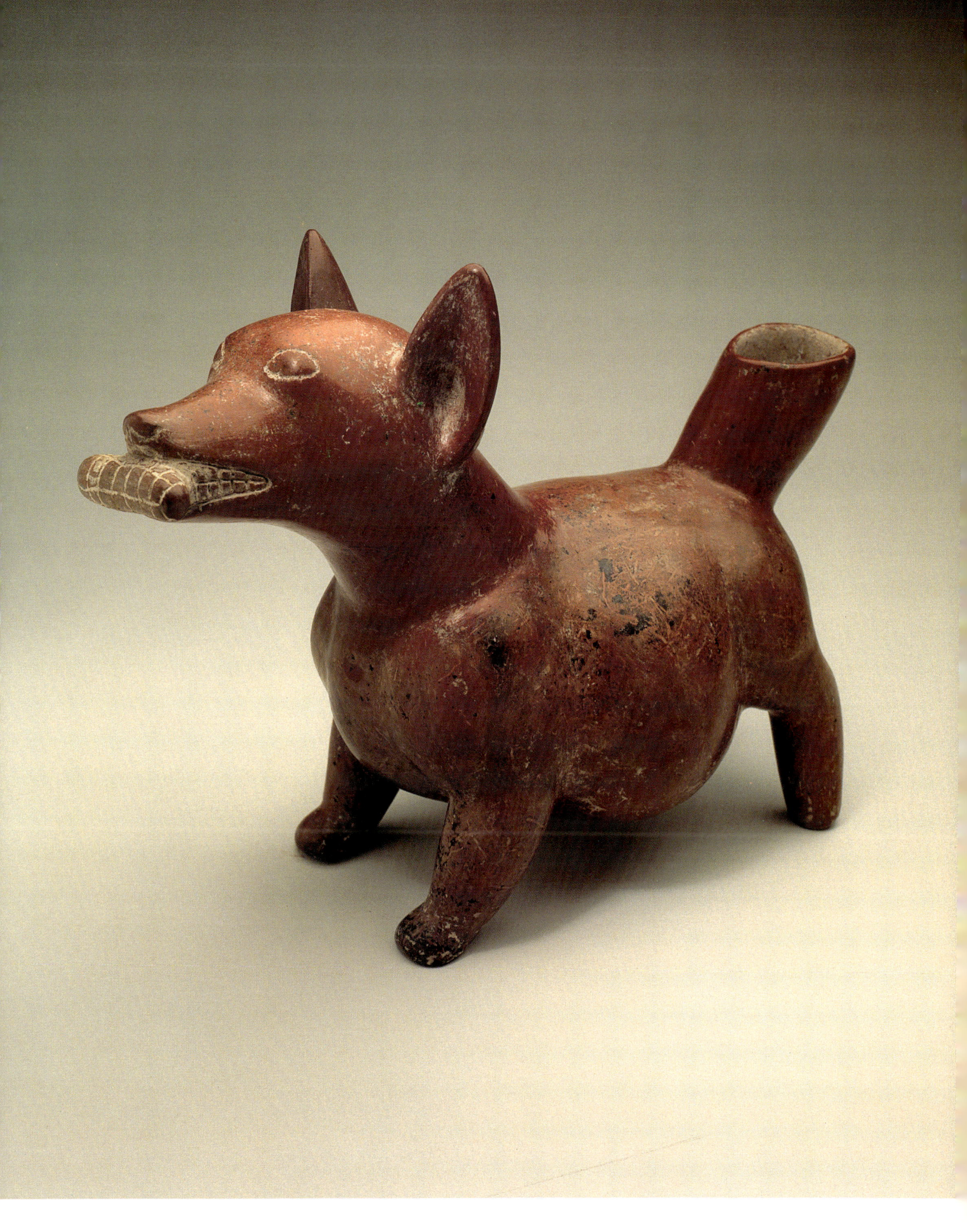

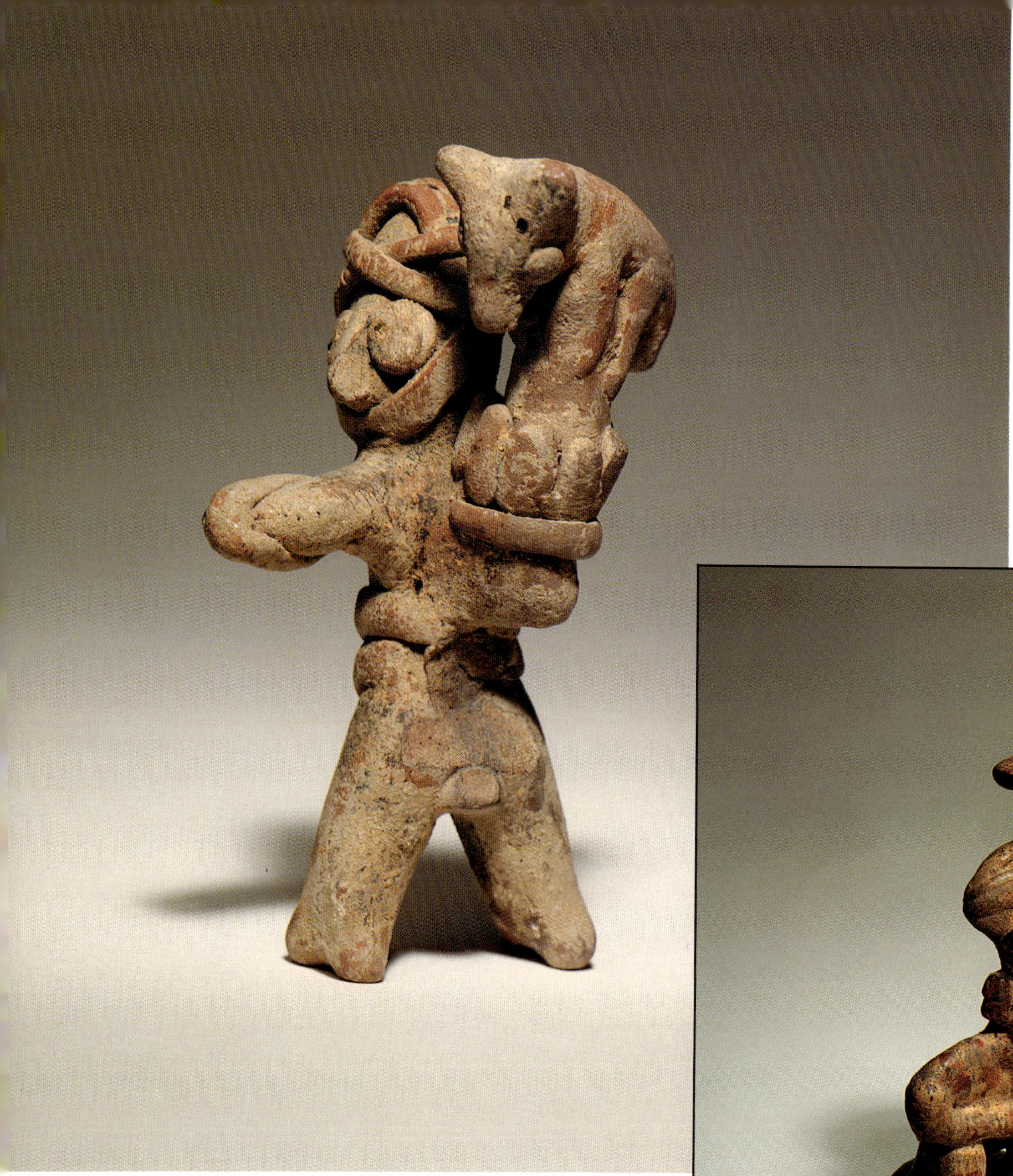

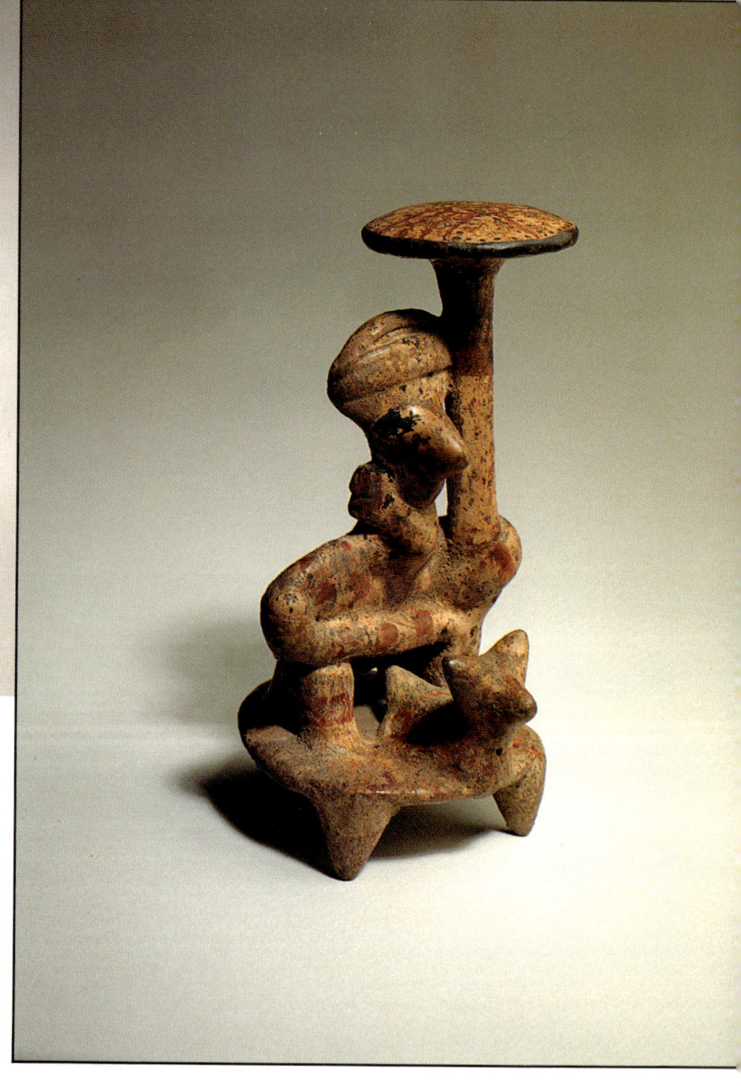

WARRIOR WITH SLING AND DOG   46
Colima, West Mexico
Late Preclassic–Early Classic: 200 B.C.–300 A.D.
Buff clay. H: 5″
Private collection

FIGURE WITH DOG AND UMBRELLA   47
Nayarit, West Mexico
Protoclassic: c. A.D. 100–500
Clay with cream slip, red and black paint. H: 5½″
San Diego Museum of Man 1978.14.193

WOMAN HOLDING A DOG   48
Tlatilco, central Mexico
Mid Preclassic: 500–100 B.C.
Buff clay with traces of red & ochre slip. H: 3¹⁄₁₆″
Lent by Constance McCormick Fearing,
Photo courtesy the Los Angeles County Museum of Art
L.83.11.515

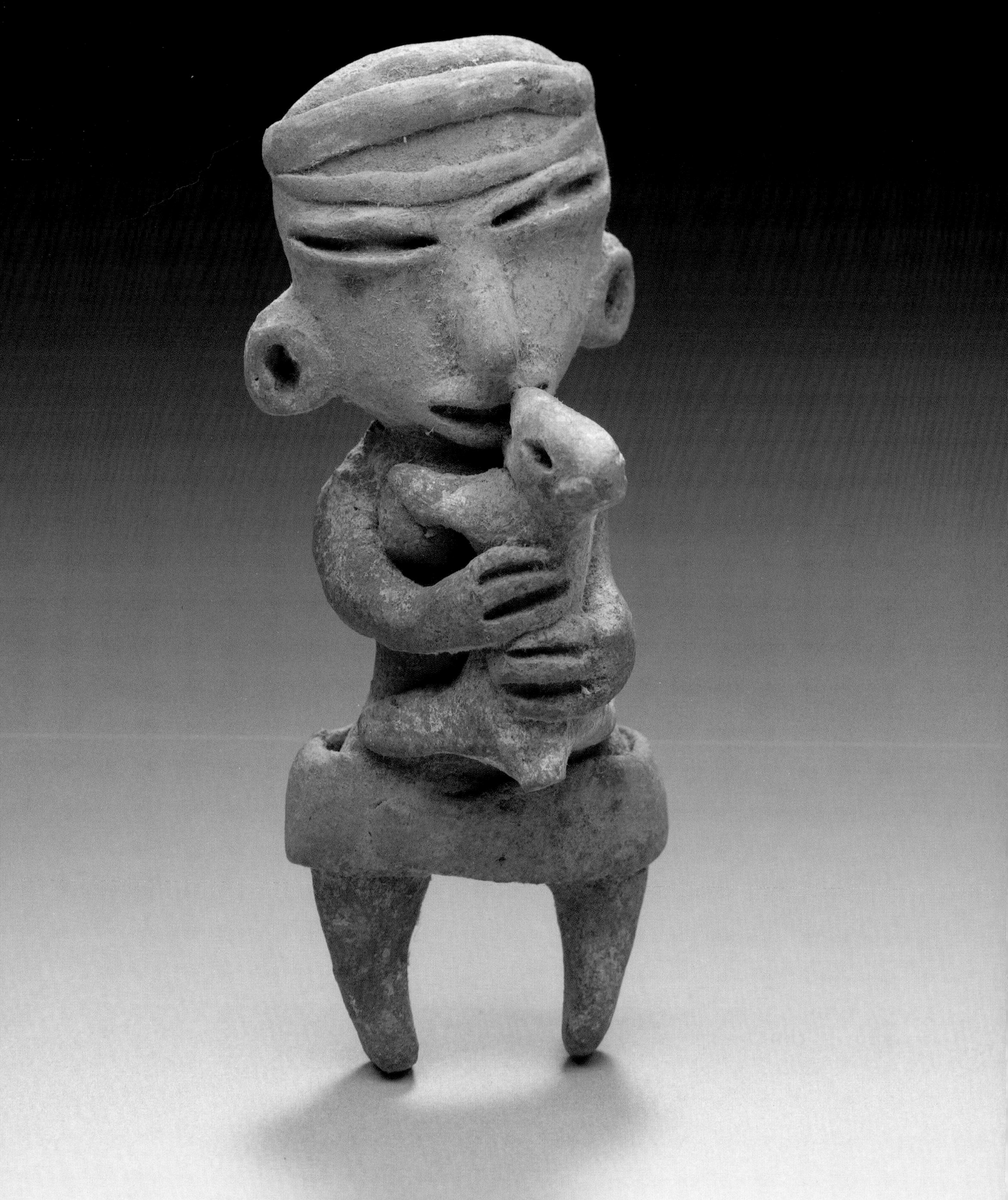

UNDERWORLD

Beneath the face of the earth lay a dark and watery realm. This underworld was at once an aquatic environment and the starry night sky. It was feared as the ultimate repository for the dead yet also perceived as a locus for the renewal of life. Named Xibalba, ''Place of Fright,'' by the Quiché Maya, the underworld was characterized by the revolting odors of rotten flesh and blood where souls were subjected to blackness, frigid temperatures, and razor sharp pains. Two of the archetypal underworld creatures, the jaguar and the bat, inflicted further tortures. A similarly dank and forbidding Underworld awaited those who met an ordinary death among the Aztecs. Mictlan or ''Place of death'' was a destination of no exit and no return.

Passage to the underworld was made through bodies of water or fissures in the earth, such as canyons or caverns. This entrance was conceptualized as the yawning mouth or skeletal maw of the earth monster and visually recreated in the architectural designs of some temple doorways.

The Classic Maya left us a vast pictorial record of the denizens who populated Xibalba. The landscapes of surreal fantasy painted on Maya clay vessels and plates reveal an array of supernaturals, ranging from the decidedly human to the zoologically bizarre. Beasts of every ilk share and trade features. Aquatic and nocturnal animals were the principal creatures of Xibalba, although the peccary and stinging insects, such as spiders and scorpions (Fig. 77), also inhabited the underworld. As the watery surface of the underworld reflected the sky, so animals associated with the upper strata reappeared in Xibalba, such as the Principal Bird Deity, vulture, turtle, deer and monkey already discussed.

AQUATIC FLORA AND FAUNA
WATERLILY — NINFACEA   *atatapacatl; naab*

Waterlilies and waterbirds, highly visible in the Mesoamerican lowlands, became symbols for the watery surface of the underworld in Classic Maya iconography.[18] The waterlily (Nymphaea ampla) was particularly thick in the agricultural canals cut between raised fields. From the canals the Maya derived rich nutrients to fertilize crops such as corn and cotton. These waterways were also the breeding grounds for a steady supply of fish. In addition to demarcating one of the cosmic spheres, the waterlily was a significant metaphor for food and abundance as well as an elite symbol for those who maintained control over reclaimed swamp agriculture. The waterlily with a nibbling fish became a headdress element that indicated the wearer's noble status or his aristocratic profession in the arts of writing and painting.

WATERBIRDS

Along with waterlilies and reeds, all waterfowl defined the watery context that ensured sustenance. For the Aztecs, waterbirds were one of many edible gifts that flourished in their capital's lake setting. They were a source of meat and of the ornamental feathers used on ceremonial gear. At the sixth-month celebration to the Aztec rain god, Tlaloc, his priests entered the lake and mimicked the splashing and sounds of ducks, gulls, and herons. A waterbird can fly and swim and also dive beneath the water. As a bird who can penetrate all three strata, it had great cosmological significance.

In Classic Maya imagery, a long-necked generic waterbird (that combined features of a cormorant and heron) is frequently depicted on the lids of vessels (Fig. 49). Waterbirds are shown holding their catch and surrounded by water symbols indicative of the underworld (Fig. 25). Waterbird imagery also forms a part of the regalia of specific underworld deities (Fig. 50). When a Maya king wore an elaborate headdress that displayed a waterbird and fish, he was publically proclaiming his identity with the water-related Chac Xib Chac, guise of god G1.

DUCK — PATO   *canauhtli; chiichilha*

Ducks were also sacred for they too dive under water and are capable of flying long distances. The Huichol gods assumed duck shapes to travel into the depths of the great Western sea. Duck effigies from the archeological cultures of West Mexico represent with sensitivity the different species of the Anatidae family. Like Colima parrot sculptures, these earthenware duck pots vary from single to multiple configurations (Figs. 51, 52). Buried with the dead, Colima duck effigies may have acted as spirit guardians, with implications of fertility and rebirth.[19]

VESSEL WITH WATERBIRD ON LID   49
Maya. Tikal, Guatemala
Early Classic: c. A.D. 250–450
Blackware. H: 7½"
Museo Nacional de Arqueología y Etnología, Guatemala
1115 a & b

A long-necked generic waterbird with the crest of a heron and the beak of a cormorant, was often represented on Classic Maya funerary pottery. Their powers of flight and underwater dives made waterbirds excellent symbols for animals that could mediate between several realms. The incised circles on the body of this tetrapod vessel are water symbols.

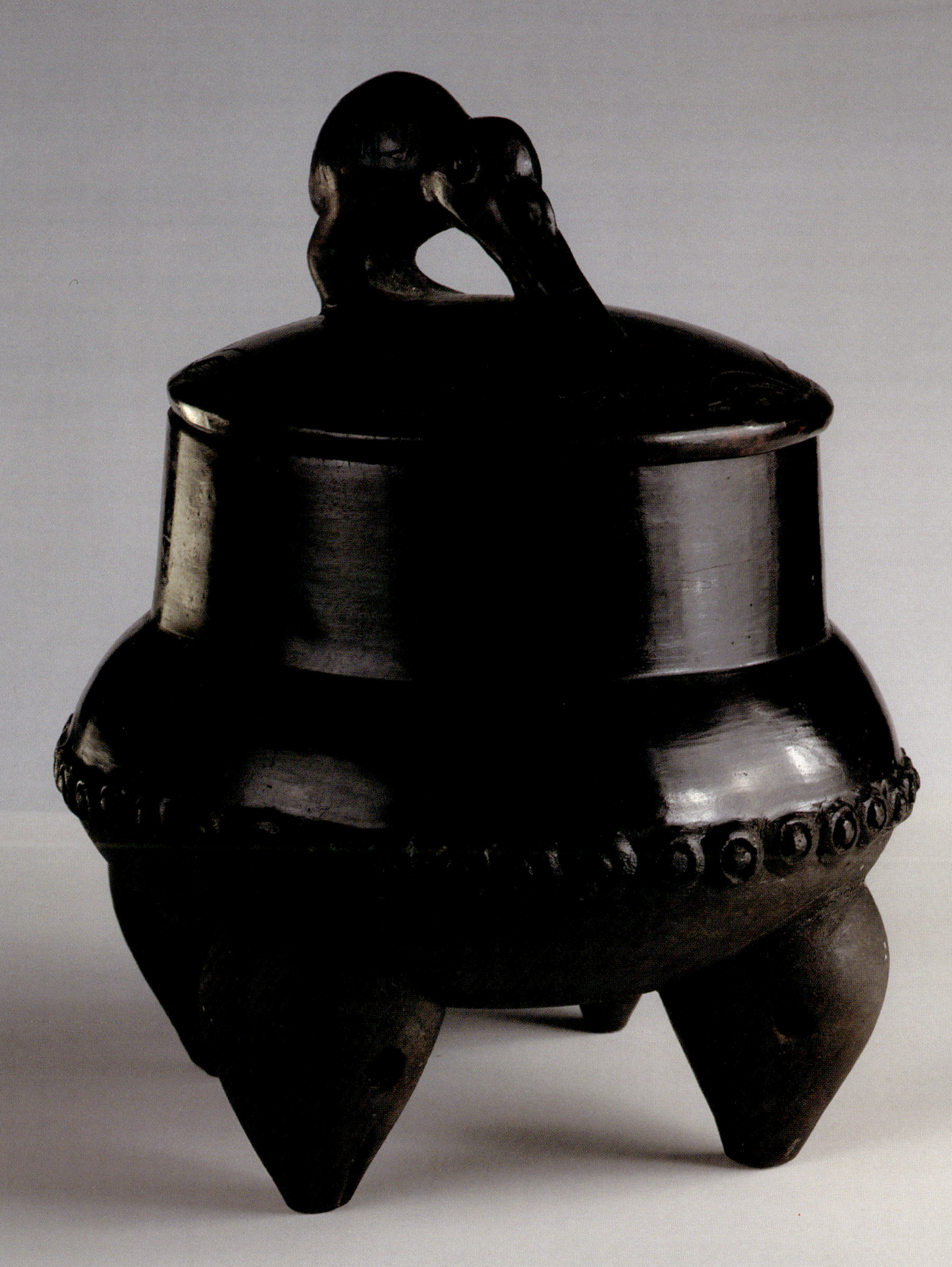

SHELLS AND FISH

Like a great subterranean sea, the underworld was home to aquatic creatures who symbolized the regenerative forces in a zone of death and rebirth (Figs. 53, 54, 55). Shells played an important role in the art and ritual of fertility cults to induce and insure rainfall. Marine and freshwater shells were traded to the highlands as part of an active exchange network. In addition to the important conch shell, a variety of clams, snails, abalone, scallop shells, pearl oysters, the spiny oyster (Spondylus) were cut and worked into ritual items and ornaments by craftsmen.

Their overriding significance as symbols of water made marine shells important in burials and in offerings to the gods. Along with red cinnabar (blood) and jade (blue-green water), Spondylus shells were buried with a Maya ruler as part of an assemblage that symbolized "precious liquid" and the king's eternal life (Coe 1988).

An impressive quantity of marine life was buried within the main Aztec Templo Mayor of Tenochtitlan. In these offertory caches, shells, sea urchins, coral, sawfish, and snails, recalled an ancient cult of fertility and water. They further extolled the political might of the Aztecs, whose conquests extended to both coasts. Most importantly, these sea symbols defined the Aztec view of the cosmos by alluding to the large body of water that encircled and flowed below the earth (Broda 1987).

CONCH SHELL — CARACOL DE LA MAR

tecciztli; hub

The bellowing, mournful sounds of the snail-like marine conch (Strombus giga) were heard at most Precolumbian rituals. The conch trumpet was formed by cutting the tip of the spire to make a mouthpiece and drilling airholes into the shell wall (Fig. 57). Among the many ceremonial occasions at which conches were blown, the Classic Maya used them to communicate with supernatural beings, often during bloodletting rites. On the shell trumpets themselves are incised the name of the user or the ancestors being called forth (Schele and Miller 1986: Pl.27). It was played with other instruments at funeral processions and sometimes then buried with the deceased. The conch shell trumpet has survived in burial contexts, along with ceramic flutes (Fig. 60), ocarinas, and whistles (Fig. 27), and the turtle carapace that was struck with a deer's antler.

Most importantly, the conch shell was affiliated with the afterlife or the otherworld. As depicted on Maya polychrome ceramics, the underworld god "N" wears or inhabits the conch shell and the death god is blowing on a conch trumpet. Themes of death and rebirth given to the conch are central to the Mexican creation myth. In order to create mankind, Quetzalcoatl descended to Mictlan, the region of the dead. The Death god of the underworld required Quetzalcoatl to blow a conch trumpet to the four world directions. With the help of worms who made holes in the conch and bees whose buzz created the sound, Quetzalcoatl was able to complete his task satisfactorily. Conch shell trumpets of natural shell, and pottery replicas, are very common in the shaft tombs of ancient West Mexico (Figs. 57, 58).[20]

SHARK AND STINGRAY

Among the generic schools of fish that inhabited the underworld, the shark and stingray were singled out to symbolize its more dreaded aspects. The torpedo-shaped shark is an aggressive predator, dangerous to man himself. A shark's dorsal fin and sharp triangular teeth identified the early Olmec Fish Monster, a supernatural associated with death.

Although sharks prefer to prowl warm ocean waters, the Bull shark (Carcharhinus leucas) can travel up rivers to freshwater inland lakes. This shark was the animal prototype for the powerful and malignant Xoc demon who devoured women, children, and other animals (Jones 1985). The Xoc monster in art incorporated the shark's upturned snout and tail.

Sharks and stingrays were classified together, both rightly feared by man. When provoked or defensive, the stingray can use its dorsal spine as a lethal weapon. The stingray's spine has a cutting-sharp edge and a venom that can inflict serious if not deadly harm. Like shark's teeth, stingray spines were also highly valued as bloodletters, in rituals of self-sacrifice (Benson 1988c). The purpose of offering blood, as discussed, was to sustain the cosmos. Thus, the demonic shark and stingray embodied dual aspects of the underworld, both deadly and indirectly, life-affirming.

POLYCHROME VASE WITH   50
WATERBIRDS AND FISH
Maya. Petén, Guatemala
Late Classic: A.D. 700–900
Terracotta. H: 9½"
The Bowers Museum, Santa Ana
F81.23.1

The bodies of the two waterbirds painted on this vase form the profile face of a Maya underworld deity, toothless old god N. A skyband with celestial symbols is also included along the bottom of the vessel. This skyband often appears on Maya clay vessels buried with the dead (Carlson 1988) and may refer to the upper realm as seen from the perspective of the dead, who was buried facing up toward the sky. (In Hellmuth 1987: Figs. 351–353).

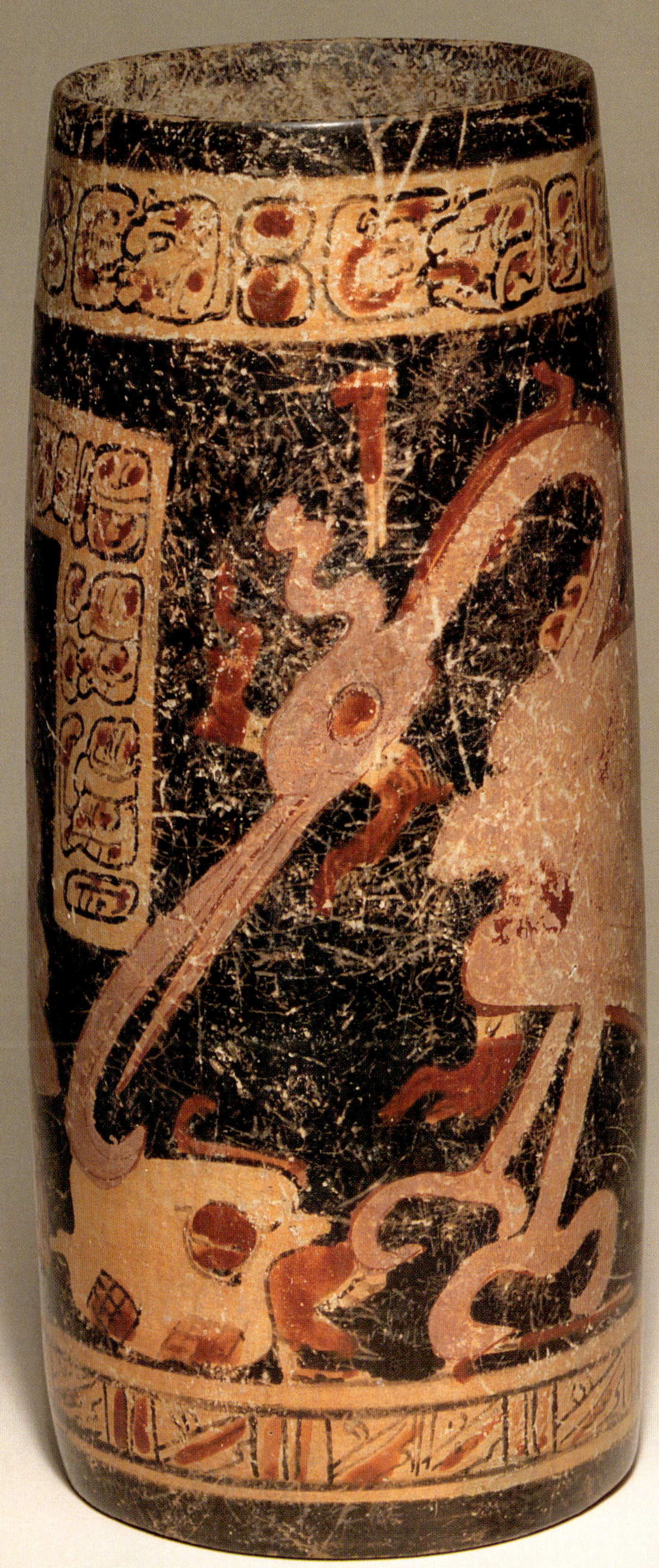

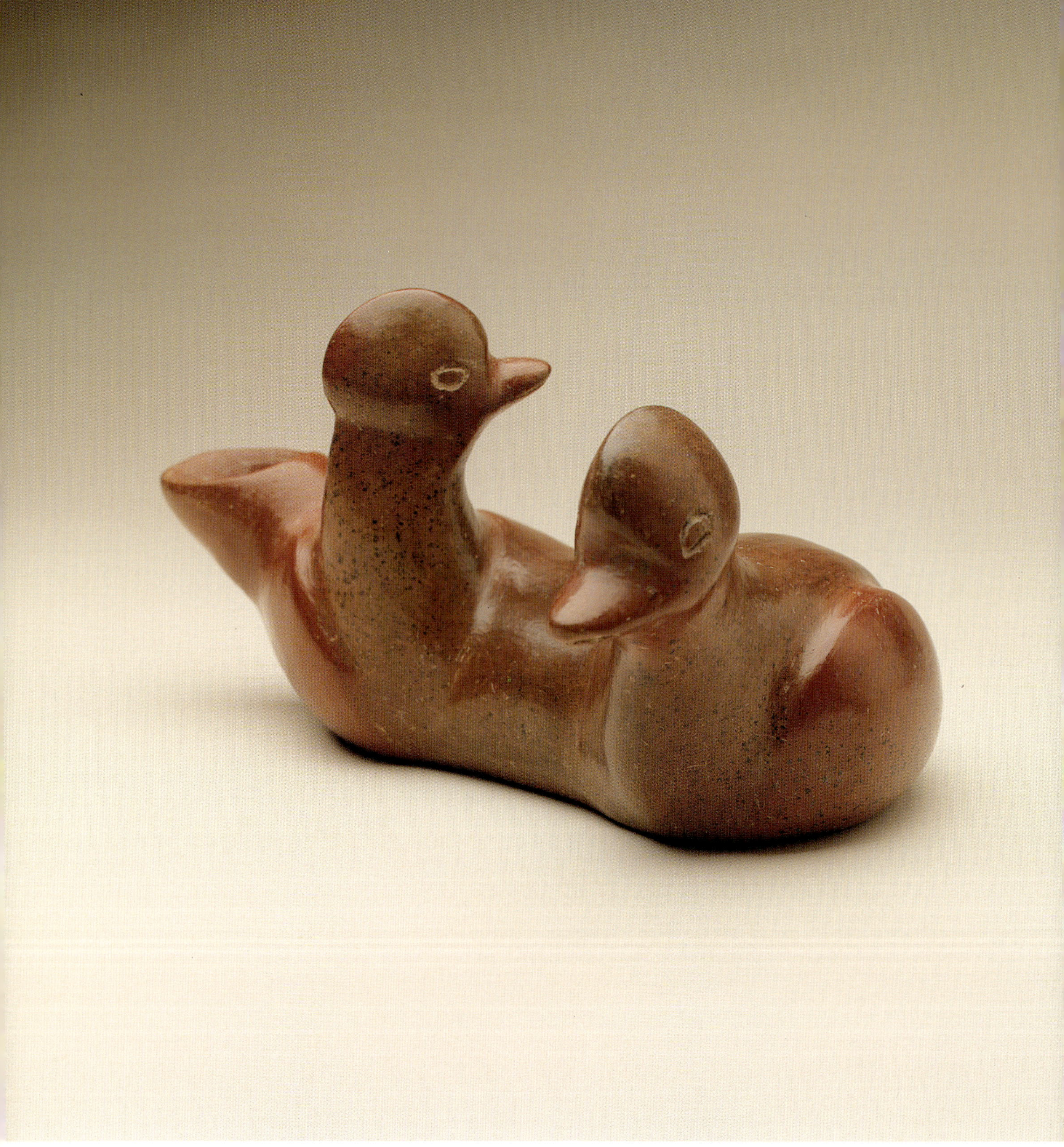

DOUBLE DUCK EFFIGY 51
Colima, West Mexico
Late Preclassic–Early Classic: c. 200 B.C.–A.D. 300
Burnished terracotta, with red and ochre slips.
H: 5¼″
Mr. William Schneider

PAIR OF JOINED DUCKS 52
Colima, West Mexico
Late Preclassic–Early Classic: c. 200 B.C.–A.D. 300
Monochrome red. H: 7½″
Frank Papworth Estate

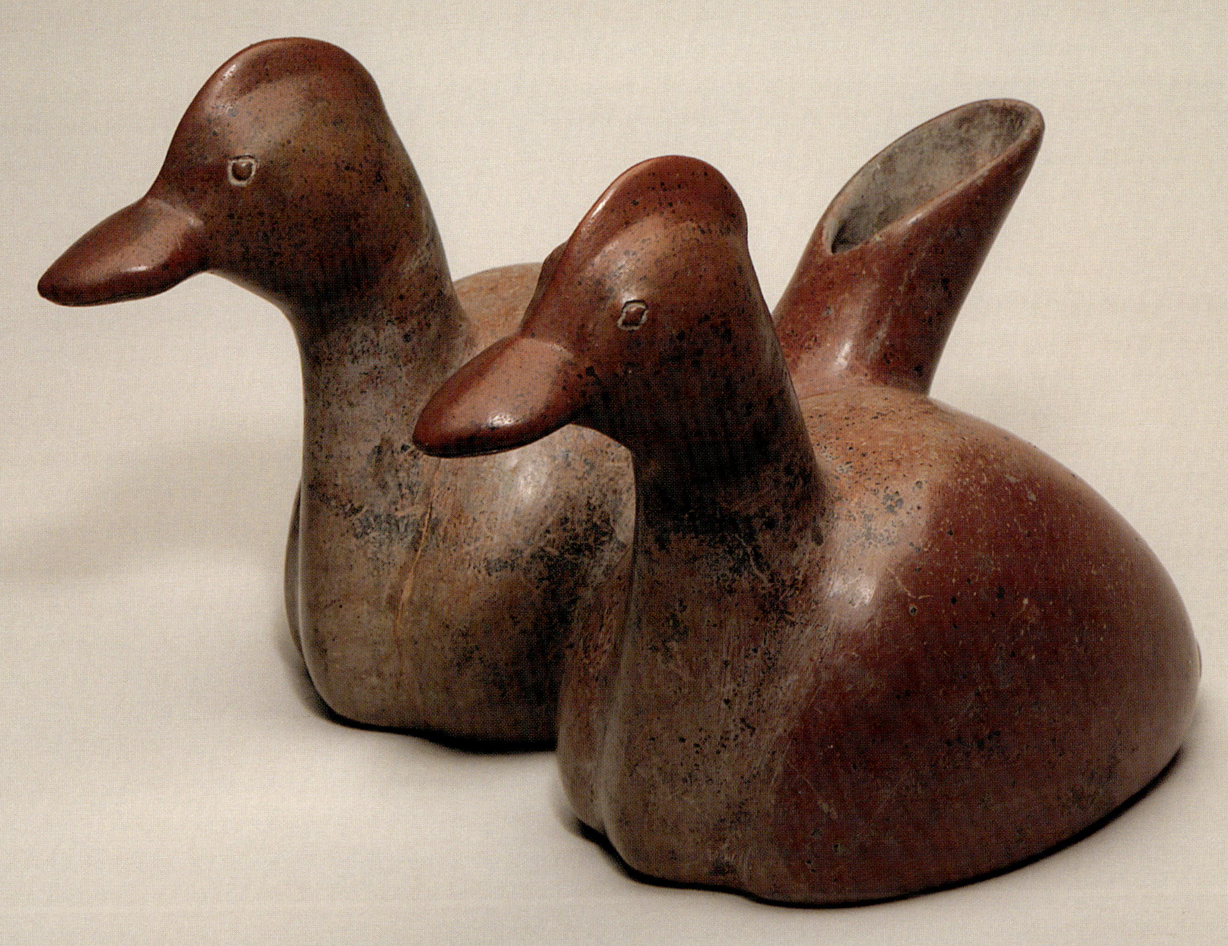

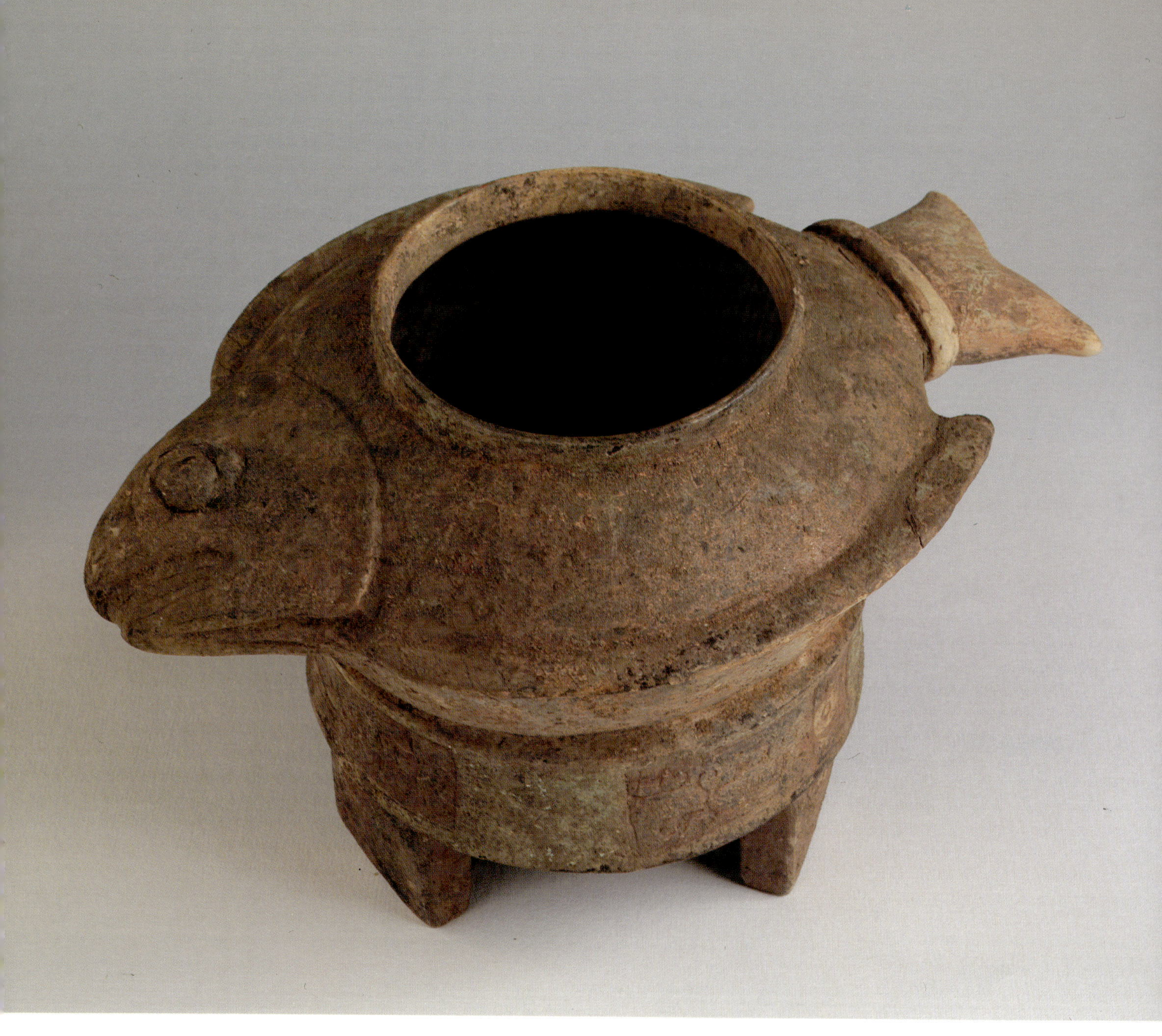

VESSEL WITH FISH EFFIGY   54
Maya. Kaminaljuyu, Guatemala
Early–Mid Classic: c. A.D. 200–550
Buff clay with remains of stucco paint. H: 7¾"
Museo Nacional de Arqueología y Etnología, Guatemala
2454

FOUR FISH ON   53
FLARING COLLAR VESSEL
Colima, West Mexico
Late Preclassic–Early Classic:
Burnished terracotta, monochrome red. D: 11"
Private collection

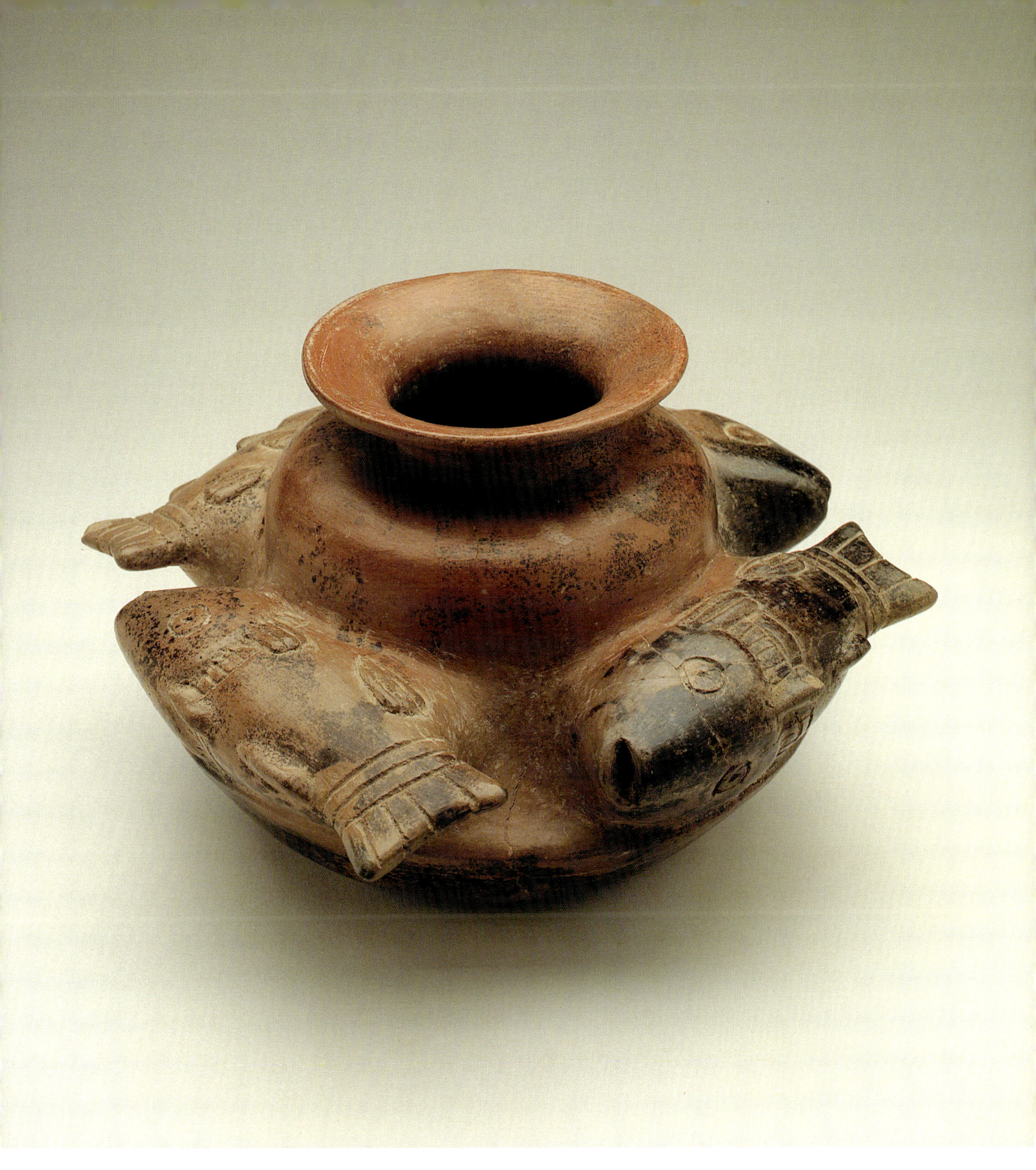

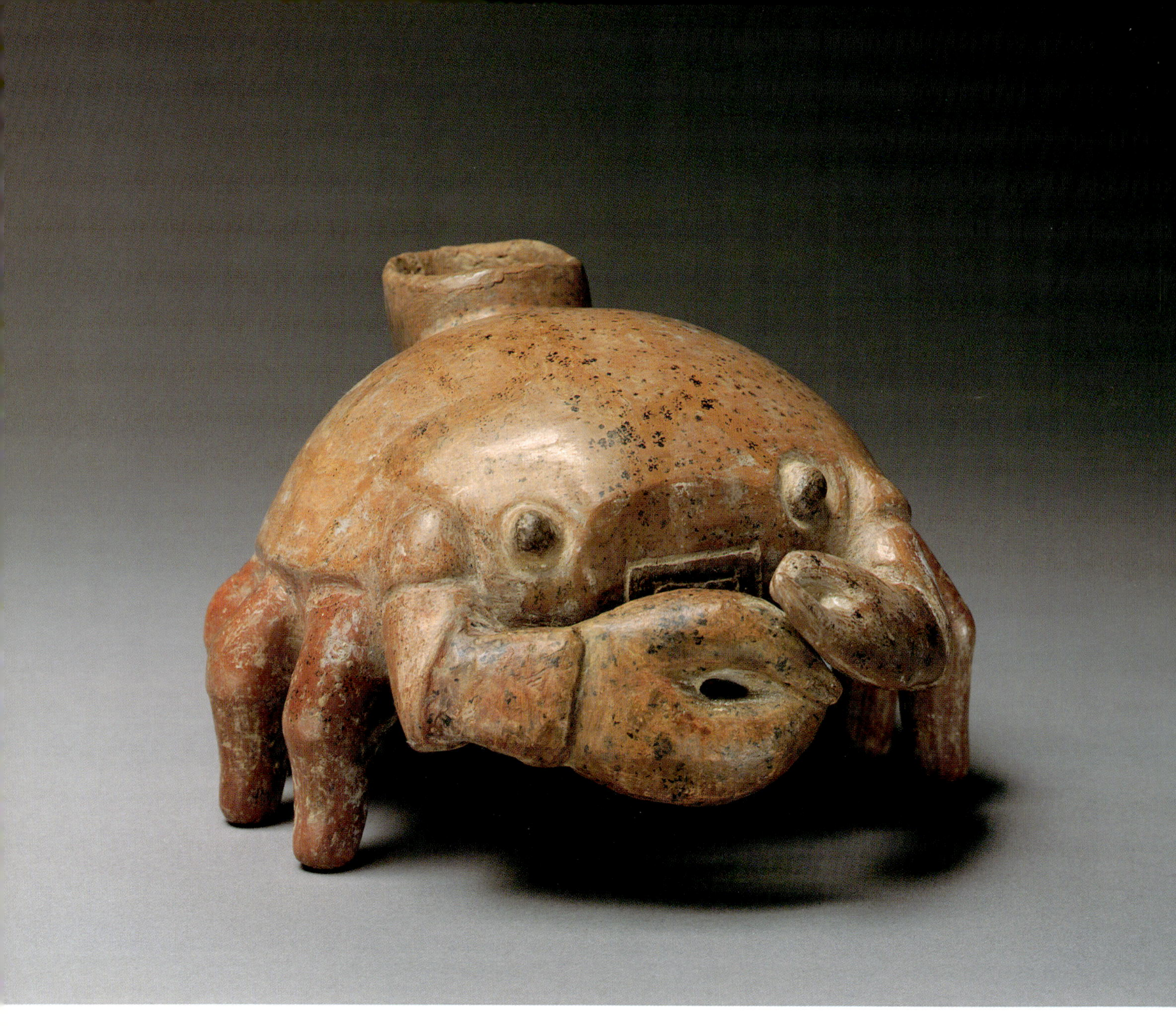

CRAB EFFIGY VESSEL 55
Colima, West Mexico
Late Preclassic–Early Classic: 200 B.C.–A.D. 300
Terracotta, burnished red slip. H: 4⅜″
Collection of The Denver Art Museum, gift of Mr. and
Mrs. Peter Natan 1974.232

**FIGURE WITH FISH
(SHARK?) HEADDRESS** 56
Colima, West Mexico
Late Preclassic–Early Classic: 200 B.C.–A.D. 250
Burnished terracotta with bichrome slips. H: 15″
Private Collection

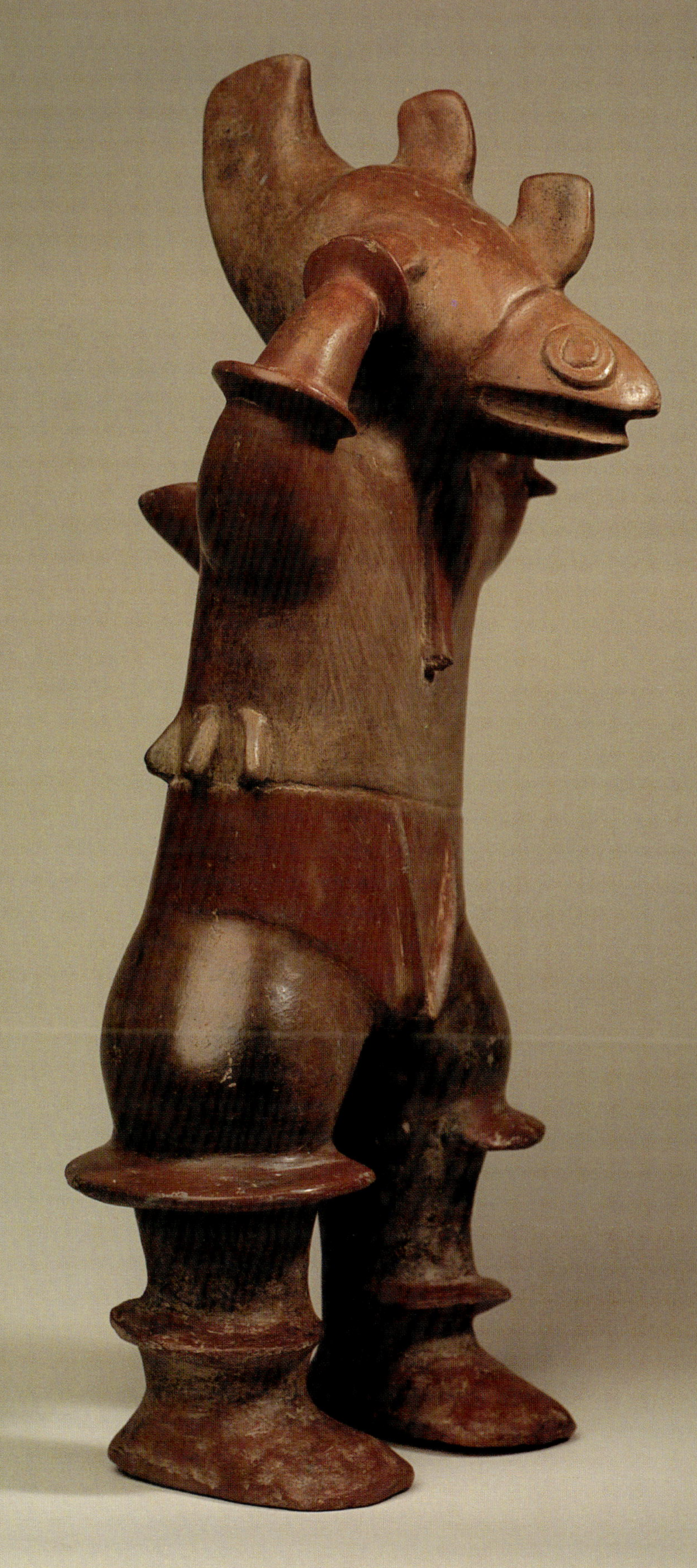

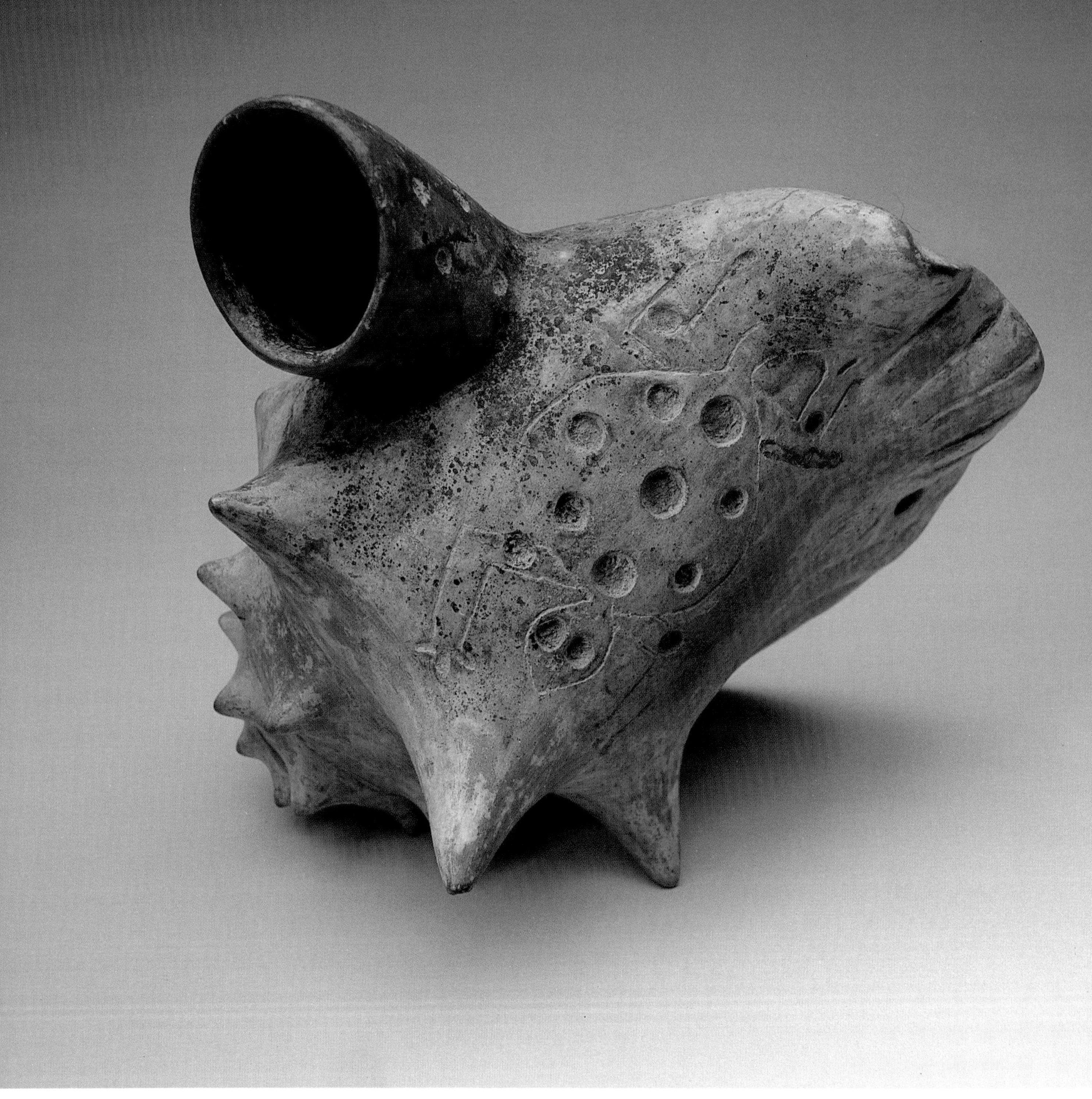

CONCH TRUMPET WITH INCISED HORNED TOAD   57

Colima, West Mexico
Late Preclassic–Early Classic: 200 B.C.–A.D. 300
Clay. 10¾″
Mr. William Schneider

Conch trumpets of natural shell and pottery replicas were frequently among the contents of shaft tombs of West Mexico. Over one hundred Caribbean conch shells were excavated from a single Colima tomb (Furst 1978:13). Traded across Mesoamerica, they testified to the deceased's wealth and status. Conch shells were

particularly affiliated with the afterlife. The incised horned toad design reinforced the conch's underworld meaning.

FIGURE BLOWING CONCH TRUMPET   58

Colima, West Mexico
Late Preclassic–Early Classic: 200 B.C.–A.D. 250
Burnished terracotta with red and ochre slips   13¾″
Mr. William Schneider

The large opening of the conch is cradled in the player's hand when used as a trumpet. Among the Aztecs, the blowing of conch trumpets at important ceremonies, even in time of war, was the privilege of priests.

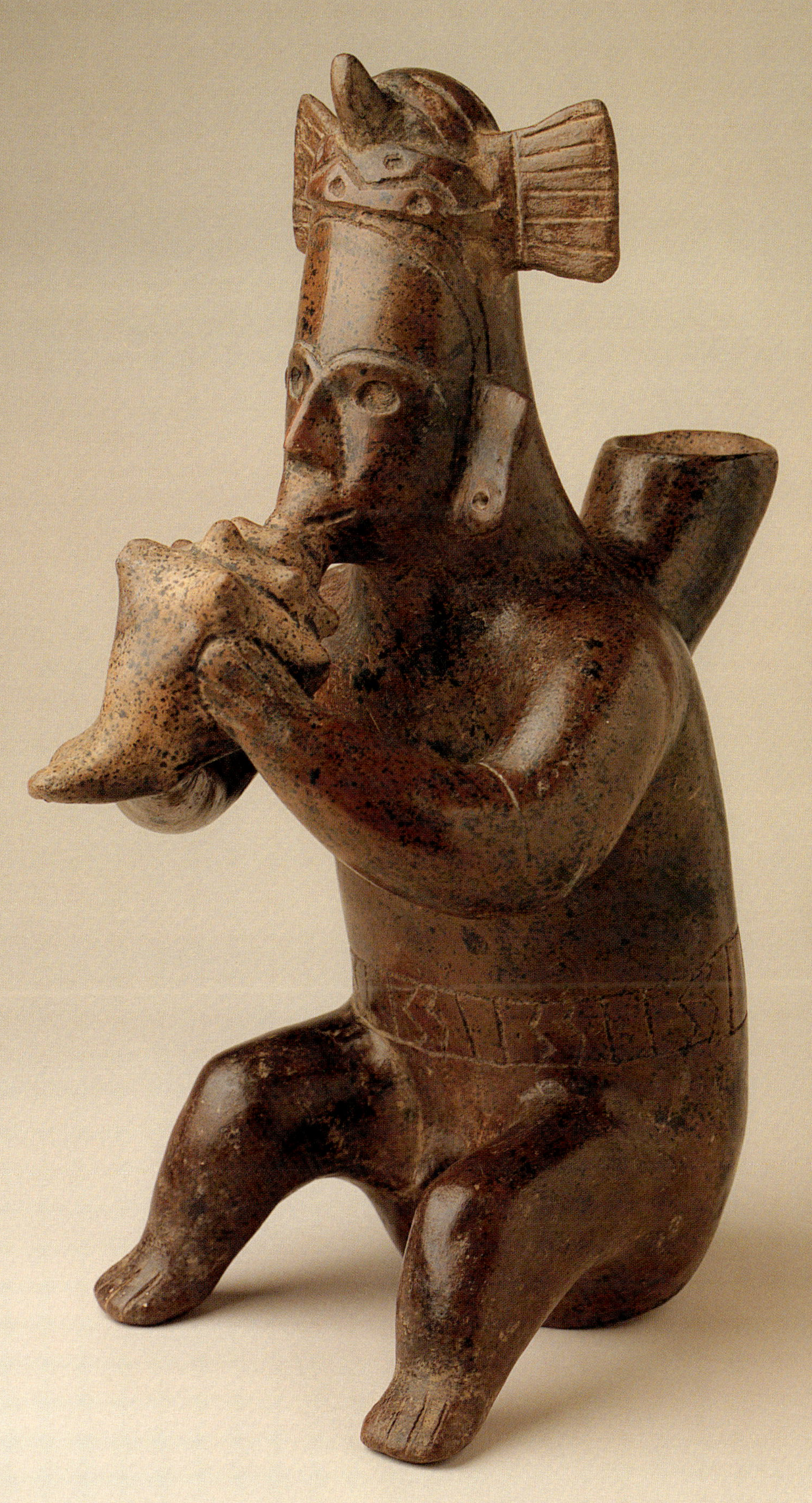

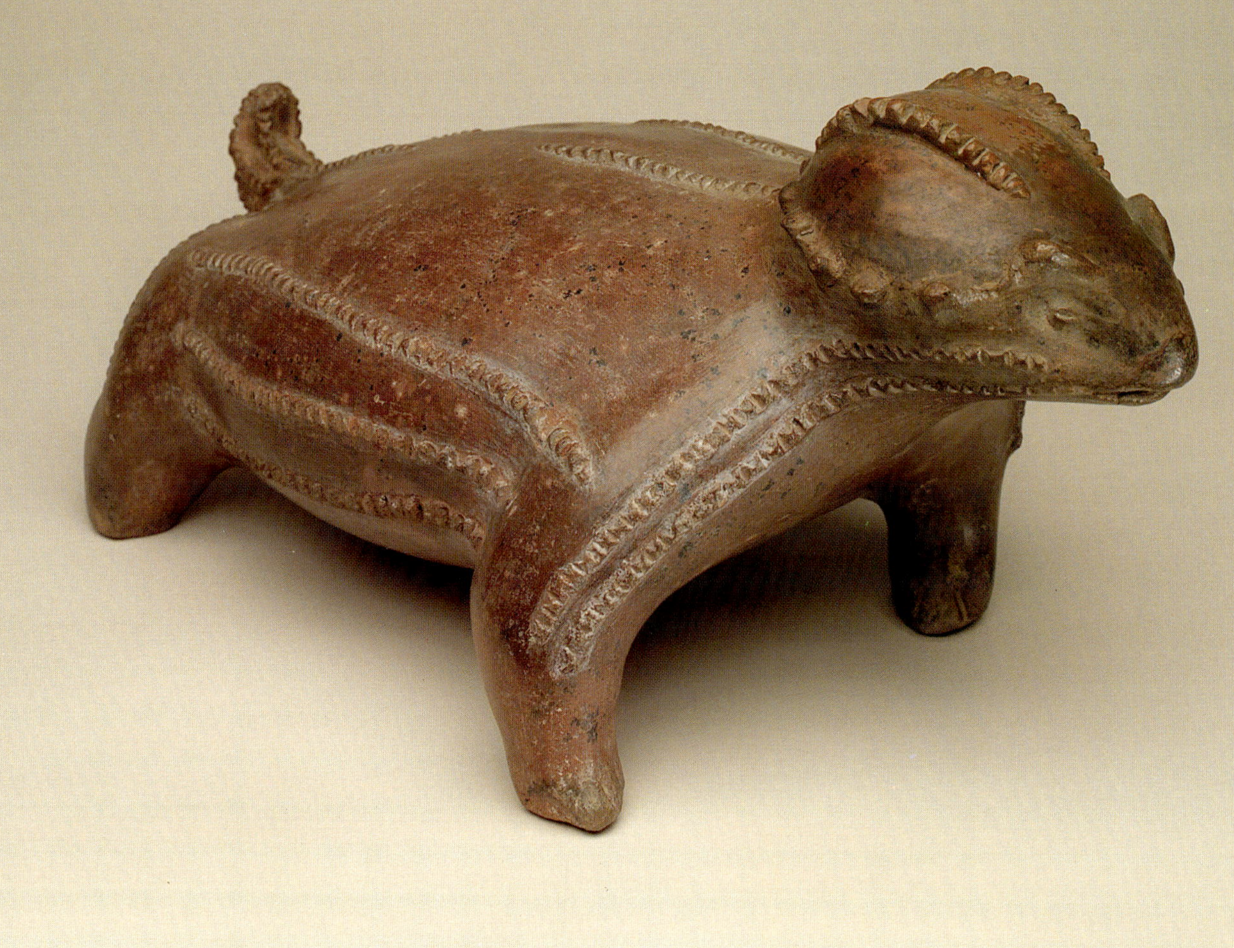

HORNED TOAD 59 Colima, West Mexico c. 200 B.C.–A.D. 300 Burnished, red-slipped terracotta. H: 8″ UCLA Fowler Museum of Cultural History, gift of Mr. and Mrs. Skip Brittenham and Mr. and Mrs. Ken Ziffren. 81.00262	**DOUBLE FLUTE WITH** 60 **ANTHROPOMORPHIC IGUANA** Veracruz, Gulf Coast, Mexico Classic: c. A.D. 200–300 Buff clay with paint traces. H: 10½″ The Land Collection ML 715 (In Nicholson & Cordy-Collins 1979: Fig.123)

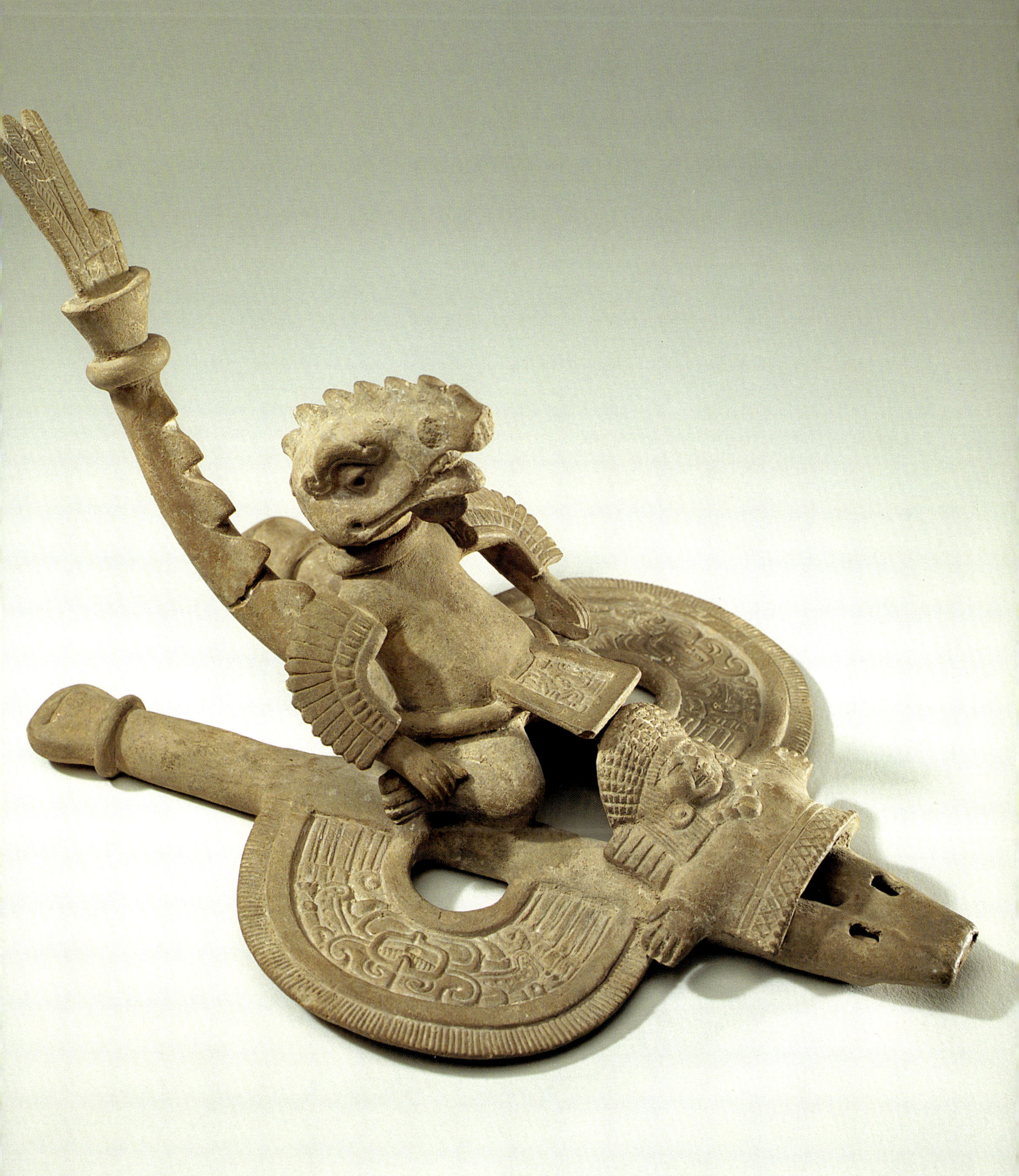

NOCTURNAL ANIMALS
JAGUAR — JAGUAR, OCELOTE *ocelotl; balam*

The King of Beasts in Mesoamerica is the jaguar (Felis onca). As the largest mammal, the nobility of this powerful spotted cat was acknowledged in the sixteenth century. "It is the lord, the ruler of the animals . . . it is noble, proud" (Sahagún 1950–1982, XI:1). The jaguar is cunning and aggressive as it stalks its prey. Its roar reverberates through the forest; sharp claws and teeth can dispatch a deer with deadly efficiency. The animal's high speeds and massive strength make it dominant in nature; they inspire man's respect and awe.

As an exclusively nocturnal hunter, the jaguar was thought of as an underworld being. The spotted black on gold hide of the jaguar was interpreted as the starry sky and helped relate the animal to the night sun.

In ancient Mesoamerica, these qualities were transferred to cultural metaphors of religious, political, and military import. Most persistent was the jaguar's affiliation with the dark interior of the earth and fertility (Figs. 61, 62). This association recognized the animal as a nighttime predator as well as a fisherman with a love of water. As early as the Olmec, earth-related supernaturals were given feline elements, particularly a snarling mouth and fangs. Overt fertility was expressed by the Olmec werejaguar, an infant figure with a down-turned mouth (Fig. 63). A jaguar monster mouth likewise characterizes the cave-like opening to the Olmec underworld.

The jaguar's reputed fertility was combined with a strong sacrificial role in the underworld (Fig. 64). These connotations shaped the complex jaguarian aspect of the Maya deity GIII of the Palenque Triad (Schele and Miller 1986: 50–51). The manifestations of GIII included a Waterlily Jaguar, the Baby jaguar with a feline tail, and the Jaguar God of the Underworld related to ritual sacrifice and war. Because of their value, jaguars were the supreme sacrificial animal. Jaguars, especially young ones, were kept in captivity, perhaps for sacrificial purposes (Fig. 65). Jaguars themselves participated in underworld scenes of human sacrifice and decapitation. These animals were frequently represented wearing the sacrificial collar, scarf, or cape, or carrying staffs of execution.

Throughout the tropical Americas, the jaguar was, and continues to be, one of the most common animals adopted by the shaman or sorcerer. The jaguar's special night vision, combined with its prowess, made it an excellent counterpart of the shaman's extraordinary powers. This spiritual identification established the jaguar as the shaman's animal familiar or nahual. As a second self, a nahual shared with man its special powers as well as protected him. Aztec conjurers and seers carried jaguar hide, tails, and claws in order to acquire the animal's daring. In addition, the shaman was often thought to transform himself into an animal during an ecstatic trance-

state. This equivalency of jaguar and shaman may be preserved in the dual meaning of the Maya word *balam*, which can be translated as both jaguar and priest. In the Postclassic period, the jaguar continued to be associated with sorcery and night. Tezcatlipoca (Smoking Mirror), the omnipotent Aztec patron of sorcerers, had multiple personalities; among his aspects was a jaguar-god of caves, Tepeyollotl (Heart of the Mountain).

Precolumbian myth and art reflect the close but ambivalent interaction of human and jaguar; the roles of hunter and hunted, assailant and victim, are often reversed (Figs. 67, 68). In the Aztec creation myth, the first era or Sun was destroyed by vicious jaguars, but thereafter, man became the master. With its reputation as destroyer, the jaguar was ultimately retained as the guardian of human society and its rulers.

Priests and rulers alike appropriated the jaguar's valor and power. Olmec, Maya, and Aztec chieftains and kings were seated on thrones decorated with jaguar imagery or covered with jaguar pelts. "He of the jaguar mat" was a title reserved for esteemed Maya officials. The spotted jaguar skin, a valued trade item, was worn as a symbol of rank and fashioned into sandals, kilts, capes, masks (Fig. 69), and full body suits (Benson 1985). Although these costume accessories varied regionally, they all proclaimed the wearer's high status. A jaguar head and mat sign *(pop)* on a royal belt signified "Holder of the Jaguar throne" (Schele & Miller 1986:114) (Fig. 70).

The imperialistic Aztecs emphasized the jaguar in their cult of war. The association of the cat's hunting stealth with a successful warrior is a logical one. Although jaguar-costumed warriors appeared in Classic period battle-scenes, the Aztecs elevated the jaguar to one of two elite military orders. Eagle and jaguar knights were rewarded for their sacrificial courage with lavish honors during life and with the promise of a celestial destination after death. In a larger sense, the pairing of the jaguar and eagle defined the complementary halves of a cosmological diagram (Benson 1988b). The eagle, as the supreme day-sun animal, represented sky and air, in opposition to the night-sun jaguar, associated with the watery environs of the earth and underworld.[21]

OWLS — BUHO, TECOLOTE *tecolotl; cui*

The owl was the nocturnal bird that the Aztecs particularly feared. "If they heard it croak or hoot over the house on which it perched, they said that someone in that house would soon die" (Motolinía 1950:153). The evil sorcerer in Aztec society was named a "human owl" *(tlacate-colotl),* a man transformed into an owl to perform black magic. As the nighttime equivalent to hawks, eagles, and falcons, owls are night hunters who kill their prey with great talons. The owl's circular eyes are set frontally like man's and exaggerated by facial disks of feathers. Also diagnostic are the feathered ear tufts in two of the most mythologically important owls, the Screech owl (Otus asio) and Great horned owl (Bubo virginianus) (Fig. 72). Native artists presented the owl frontally, thus associating it with concepts of the earth, darkness, and death (Klein 1976). Owls were and are bad omens; Maya peoples today interpret an owl's hoot as a "warning from Xibalba" (D. Tedlock 1985:272).

Indeed, the owl was the messenger used by the Underworld Lords. In Classic Maya vase painting, the toothless, wrinkled god L presided over Xibalba. Old god L sports a mythological Moan Bird in his headdress that is at least part owl. This mythological Moan Bird is readily recognized by black tipped and spotted feathers (Fig. 73). As a metaphor of death, this owl creature was a suitable icon for a work of art buried with the dead.[22]

BAT — MURCIELAGO

quimichpatlan, tzinacan, zotz

The nocturnal bat was as ominous as the owl, but far more ambivalent. As its Nahuatl name implies, the bat is a "rat that flies" *(quimichpatlan).* It has a furry, mouse-like body and a face like a cat. Yet, it is the only winged mammal that flies. Navigating by a kind of sonar, bats are aggressive, agile, and extremely adaptable; they are insectivores, fruit-eaters, flesh-eaters, fishers, and even blood-suckers. A bat is consistently in darkness; it feeds by night and roosts in caverns and shadowy overhangs, where it sleeps upside-down. The blackness and putrefying odor of a bat cave made it a fitting underworld metaphor.

Of the hundreds of species that swoop and dart in the Mesoamerican night sky, the two most important types in Precolumbian art and myth were the leaf-nosed bats (Phyllostometidae) and the vampire bats (Desmodinae). Precolumbian images often combined the pointed extension of flesh above the "leaf nose" with the sharp canine teeth of the vampire. In the Popul Vuh, Xibalba's bats are "monstruous beasts, their snouts like knives, the instruments of death" (Tedlock 1985:143). The vampire bat's diet of living blood further linked the animal to blood sacrifice. On Classic Maya ceramics, "Killer bats" have eye balls and crossed bones on their wings and hold sacrificial bowls. The blood-sucking bat continues to manifest itself as a likely originator of the Tzotzil Blackman, a winged black demon (Blaffer 1972). The Blackman of the Mayan community of Zinacantan is sinister and vengeful; he is considered a cannibal and sexual molester. His sexual potency is displayed in a huge, death-dealing male organ.

Sexual connotations arose from the Precolumbian bat's connections with rain and fertility. A bat symbolized the dark thunderclouds of an impending rainstorm. Bat guano is excellent fertilizer, but more importantly, bats are pollinators of flowers. As they fly from one night bloomer to the next, bats not only suck nectar with their long tongues but collect and spread pollen. Significantly, one of the night bloomers is the Ceiba (Silk-cotton tree), the World Tree of the Maya also associated with rain and ancestry. Blaffer (1972:66) notes that as a possible symbol of their potency, bats enter caves alone but always appear to emerge in large swarms.

Closely associated with fertility and maize production, bats figure prominently in the late Preclassic and Classic sculpture of the Zapotecs in Oaxaca (Fig. 74). Among the other bat-related deities in the Zapotec pantheon, the most important is the corn god, Pitao Cozobi (god of Glyph L), whose headdress includes a bat head, cobs of maize, and flowers.[23]

TAPIR — TAPIR, DANTA

tlacaxolotl; tzimin

The tapir (Tapirus bairdii) competes with the jaguar as the largest quadruped of the tropical forests. Now almost extinct in Mesoamerica, this curious animal is the size of a small pony and can weigh up to 660 pounds. Shy and myopic, it is nonetheless remarkably fast moving for its size and weight. The tapir's most distinctive feature is its long downcurving snout and upper lip which the animal can use like a rudimentary trunk when it feeds on leaves. The tapir emerges primarily at night to eat its vegetarian diet and swims well. In fact, its need of water restricts the tapir's habitat to the estuaries and lagoons of the lowlands. Known primarily by reputation in the central plateau, the Aztecs nonetheless had a name for the tapir, that of "monster man" *(tlacaxolotl).* The Maya associated the tapir with the nocturnal and aquatic creatures of the underworld.

Because the tapir is a voracious eater (it can decimate an entire cornfield at a sitting) and has large genitalia, the animal was reputed to possess large appetites, both culinary and sexual. The tapir's sexual potency and its frequent immersion in water gave it strong connotations of fertility. In the sixteenth century Yucatec Indians claimed tapirs stored water in their snouts and Mayan rain deities were said to ride tapirs. In this regard, the tapir plays a role in the original creation of real human beings. According to the Cakchiquel Maya, a useful human was created when maize dough was kneaded with the blood of a tapir and serpent.[24]

PECCARY — JABALÍ

coyamatl; chitam, ac

Often paired with the tapir, the peccary also has a flexible snout and pointed hoofs. In the Popol Vuh, "Great White Peccary" and "Great White Tapir" are coupled as the oldest creator gods. Two species of these nomadic, wide-ranging animals are still fairly abundant, the Collared peccary (Tayassu tajacu) and the White-lipped peccary (Tayassu pecari). Images of the peccary are frequent in Maya art (Fig. 75), sometimes ridden by god D or Itzamna, an underworld deity.

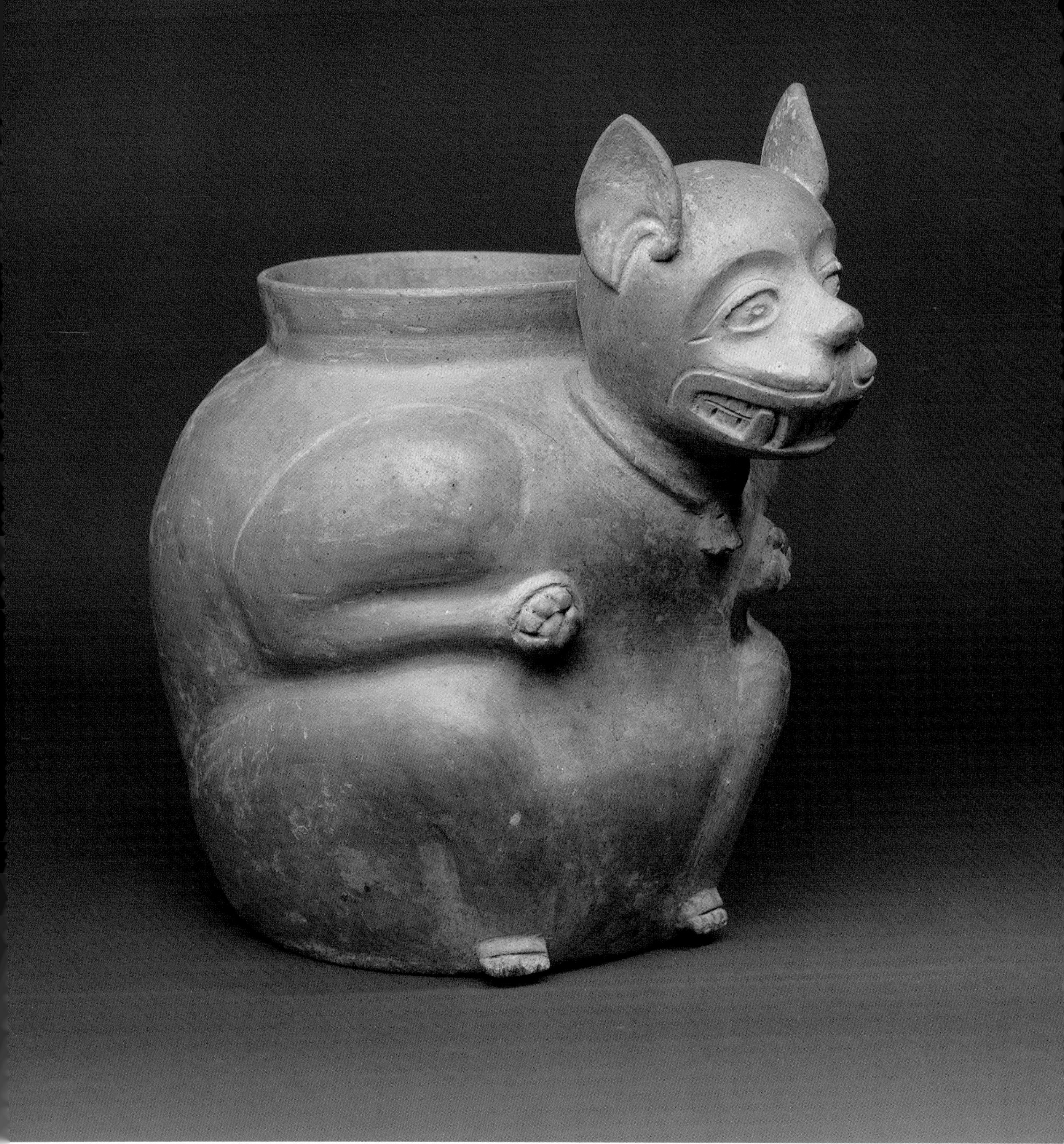

SEATED JAGUAR EFFIGY   61
Teotihuacan, central Mexico
Early Classic: c. A.D. 250–500
Thin orange ware. H: 11½"
The Natural History Museum of Los Angeles County
FA 13.75.1931

92

MASSIVE JAGUAR   62
Teotihuacan. Central Mexico.
Mid Classic (Teotihuacan III): c. A.D. 500–600
Blackened limestone. H: 15"
Lent by Constance McCormick Fearing,
Photo courtesy the Los Angeles County Museum of Art
L.83.11.1022

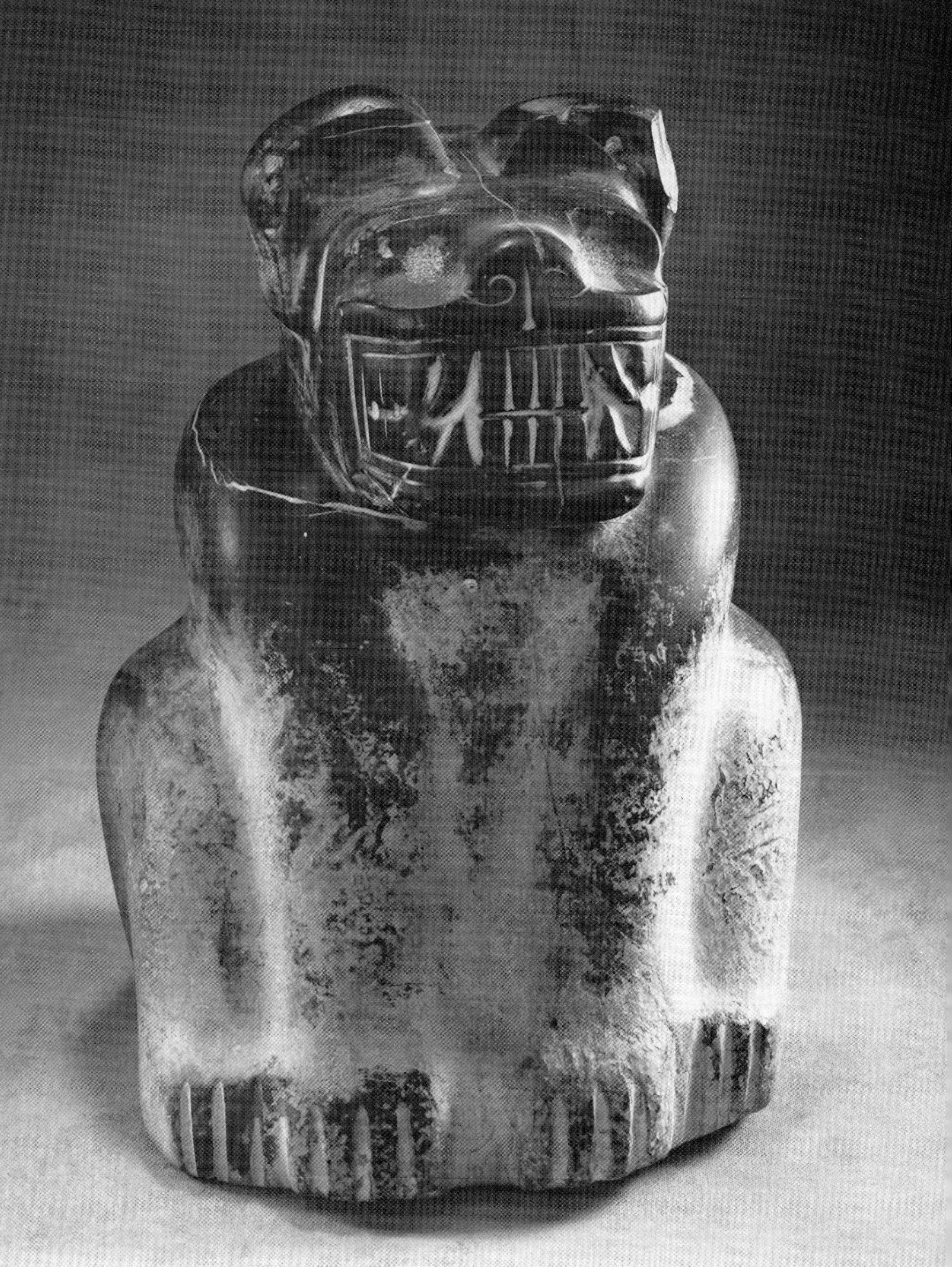

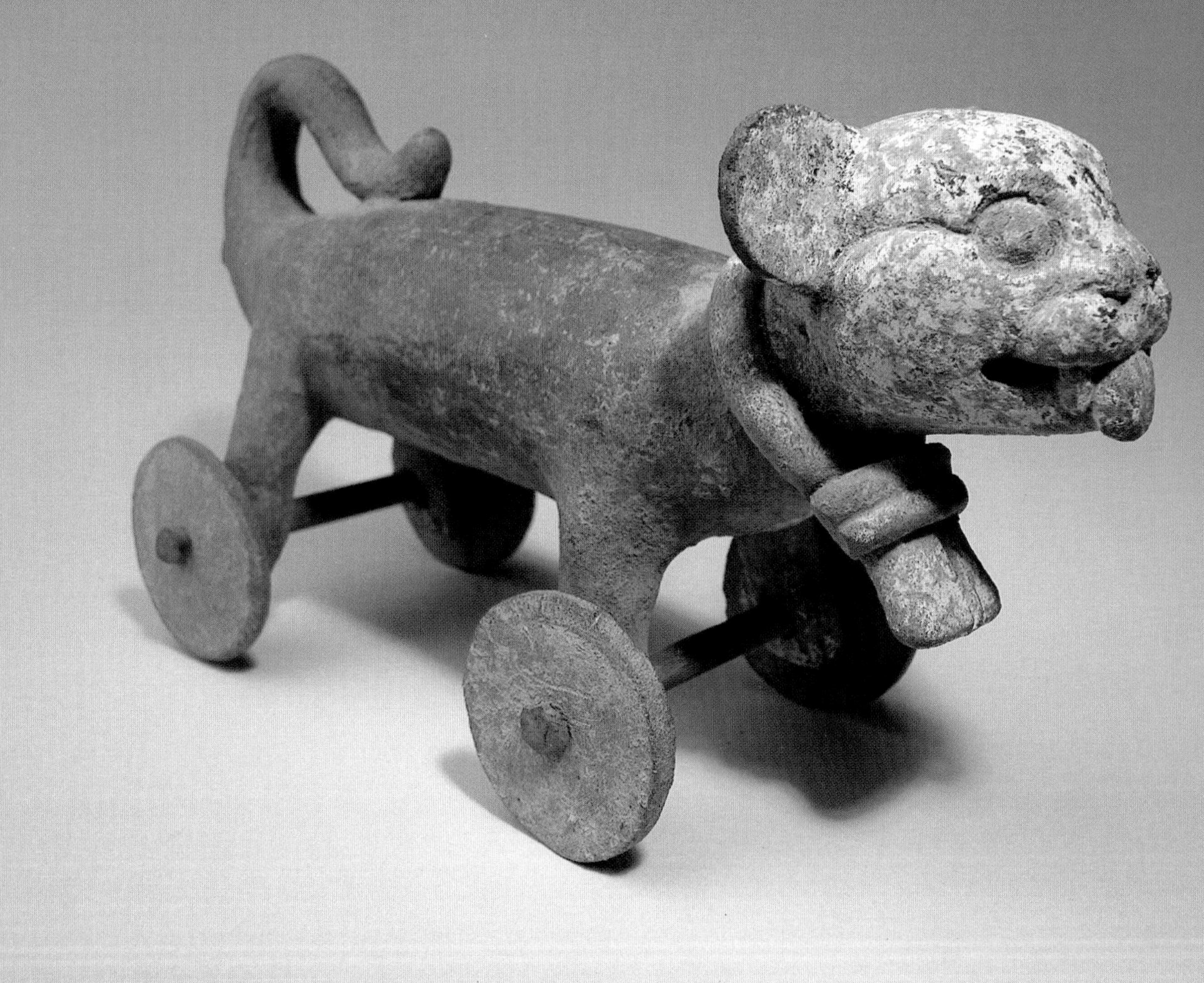

WHEELED JAGUAR   64

Veracruz, Gulf Coast, Mexico
Late Classic Veracruz: c. A.D. 750–900
Buff clay. H: 3″
Mr. William Schneider

Although the wheel was not put to practical use by New World man, perhaps because there were no draft animals, small wheeled clay figures are known from several regions of Precolumbian Mesoamerica. Wheeled dogs and jaguars are frequently found in graves and had ritualistic importance. The underworld symbolism of these animals made them suitable companions for the dead.

WERE-JAGUAR AXE   63

Olmec. Veracruz, Gulf Coast, Mexico
Mid Preclassic: c. 1100–800 B.C.
Stone. H: 11½″
Collection of The Denver Museum of Art, museum purchase with funds from Collectors' Choice VI
1985.14

The Olmec ''were-jaguar'' combined human infant with a jaguar or toad's down-turned mouth. When the ''were-jaguar'' appeared on Olmec votive axes, it had fertility significance, made clear by the maize plant symbol sprouting from the cleft in its head.

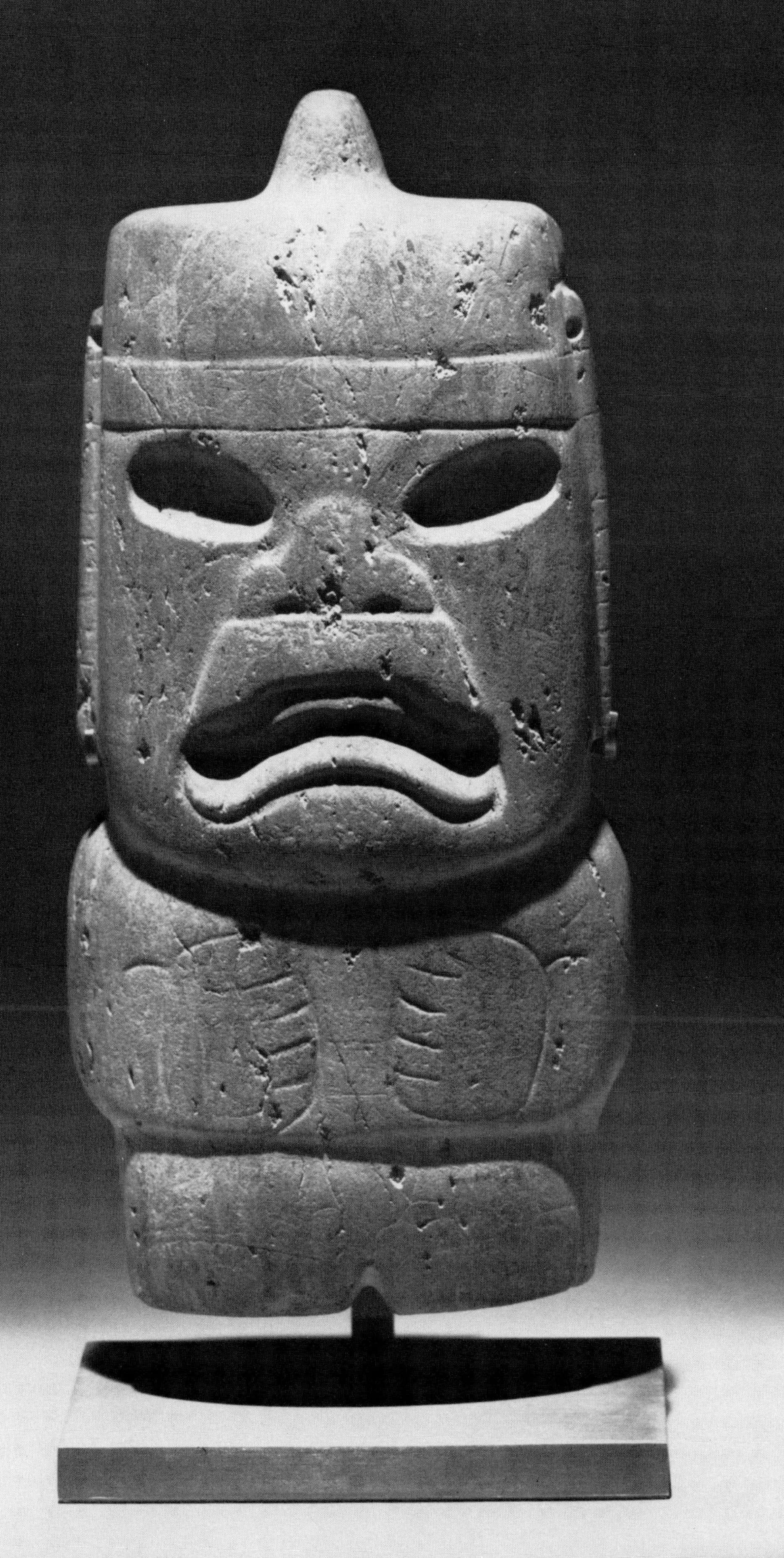

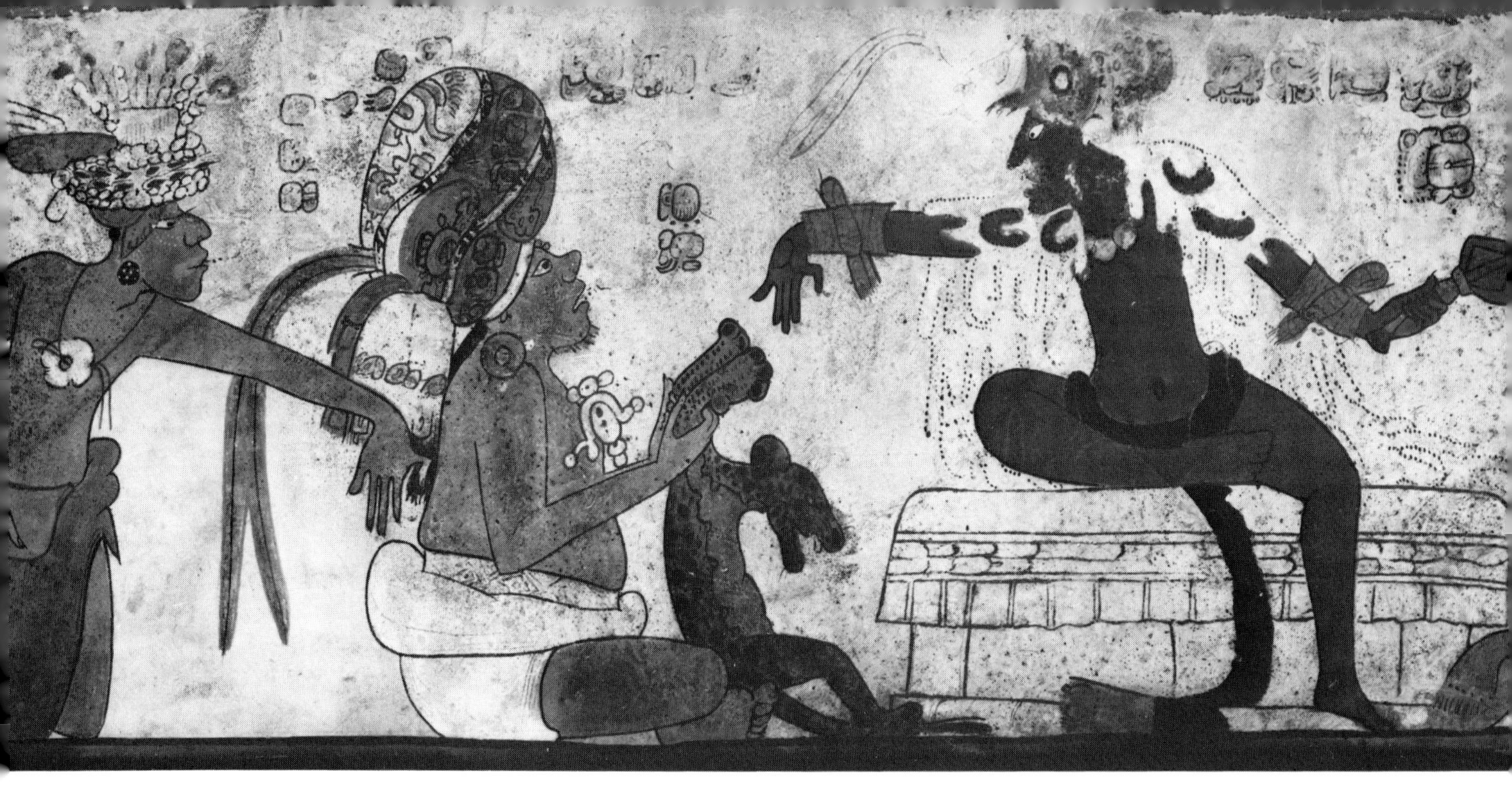

ROLLOUT PHOTOGRAPH OF FIG. 65
Photograph Justin Kerr 1990, File #2697

The scene painted on this cylinder vase from Tikal shows the presentation of a jaguar cub to a Maya lord or by a well endowed aristocrat. The lord is seated on a bench throne, wears a splendid white feathered cape, and gestures grandly. The jaguar may be a gift to the lord, to be raised in captivity for sacrificial purposes.

CYLINDER VASE WITH   65
THREE FIGURES AND JAGUAR
Maya. Tikal (Mundo Perdido), Petén, Guatemala
Late Classic: A.D. 750–800
Polychrome clay with black, tan, red, and pink on cream. H: 5¼″
Museo Nacional de Arqueología y Etnología, Guatemala 11418
(In Clancy et al. 1985: Pl. 95)

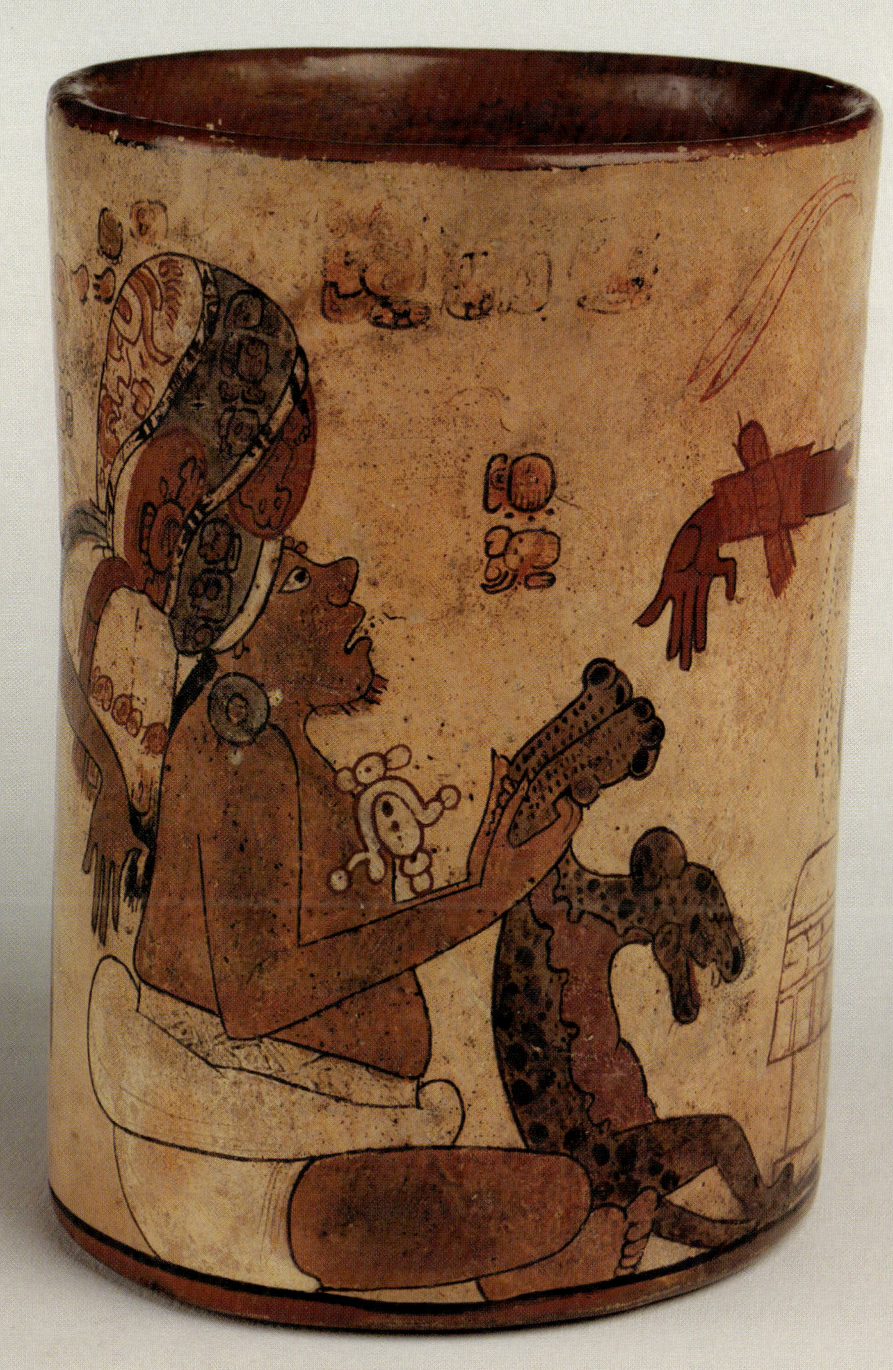

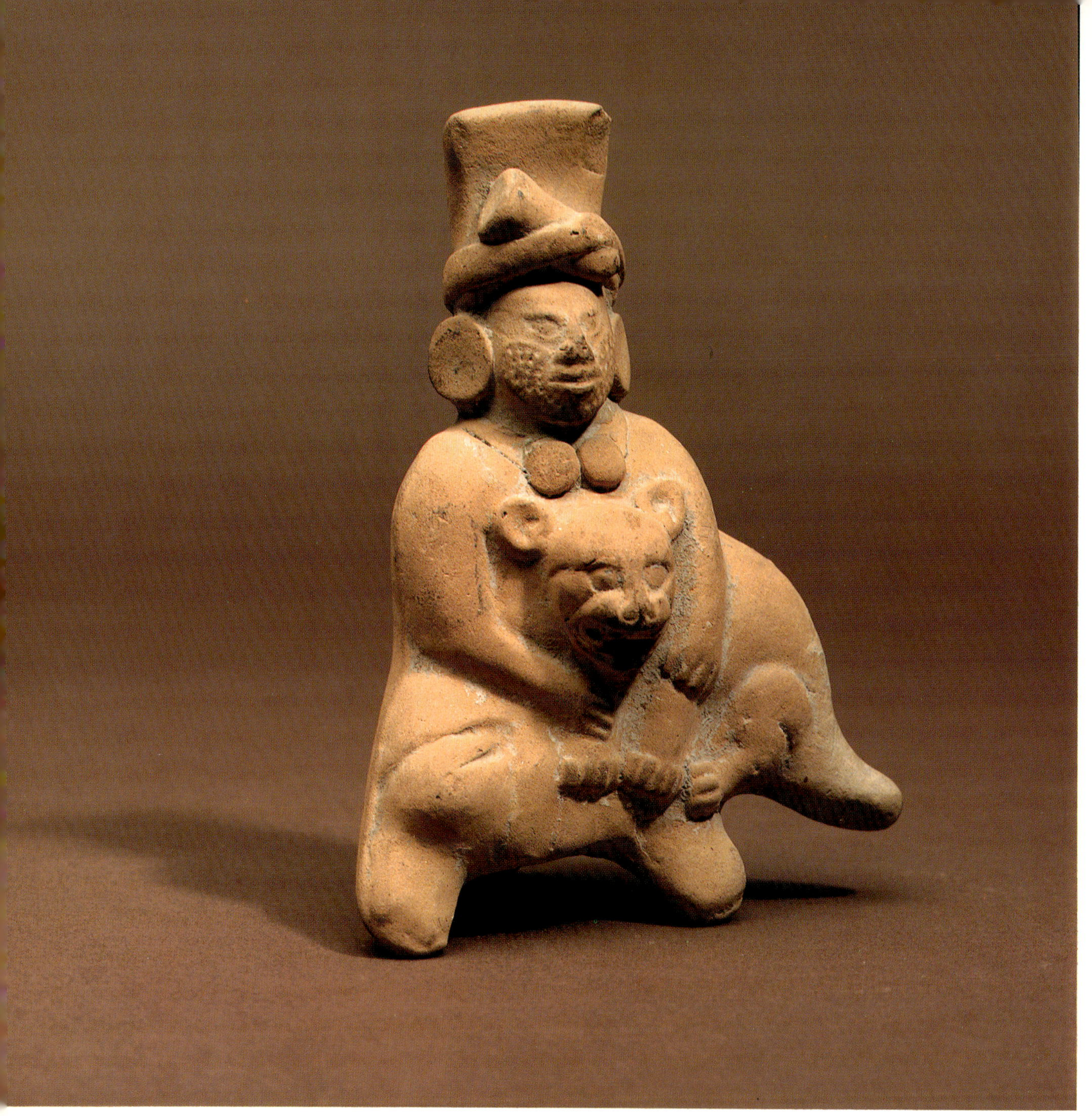

FIGURE HOLDING JAGUAR 67	**ANTHROPOMORPHIC JAGUAR** 68

FIGURE HOLDING JAGUAR 67

Maya. Jaina, Campeche
Late Classic: c. A.D. 550–800
Buff clay. H: 5″
The Natural History Museum of Los Angeles County
FA 13.74.1727

This clay whistle depicts a male figure, with padded and scarified cheeks, who is seated crosslegged. A man of rank, he is bedecked with earplugs, necklace, and turban headdress. Under his elbow he holds a jaguar, rather like a placid pussy cat.

ANTHROPOMORPHIC JAGUAR WITH HUMAN 68

Maya. Jaina, Campeche
Late Classic: c. A.D. 550–800
Buff clay, mold made. H: 4⅜″
Collection of The Denver Art Museum, anonymous gift 1983.413

In contrast with Figure 67, the jaguar in this work is dominant over man. The theme of jaguar as protector is relatively common in Jaina art, but here the animal is depicted as a devourer. The upright, anthropomorphic feline snarls to expose fangs and tongue; with its forepaws it tears open the stomach of the limp, diminutive human in its grasp.

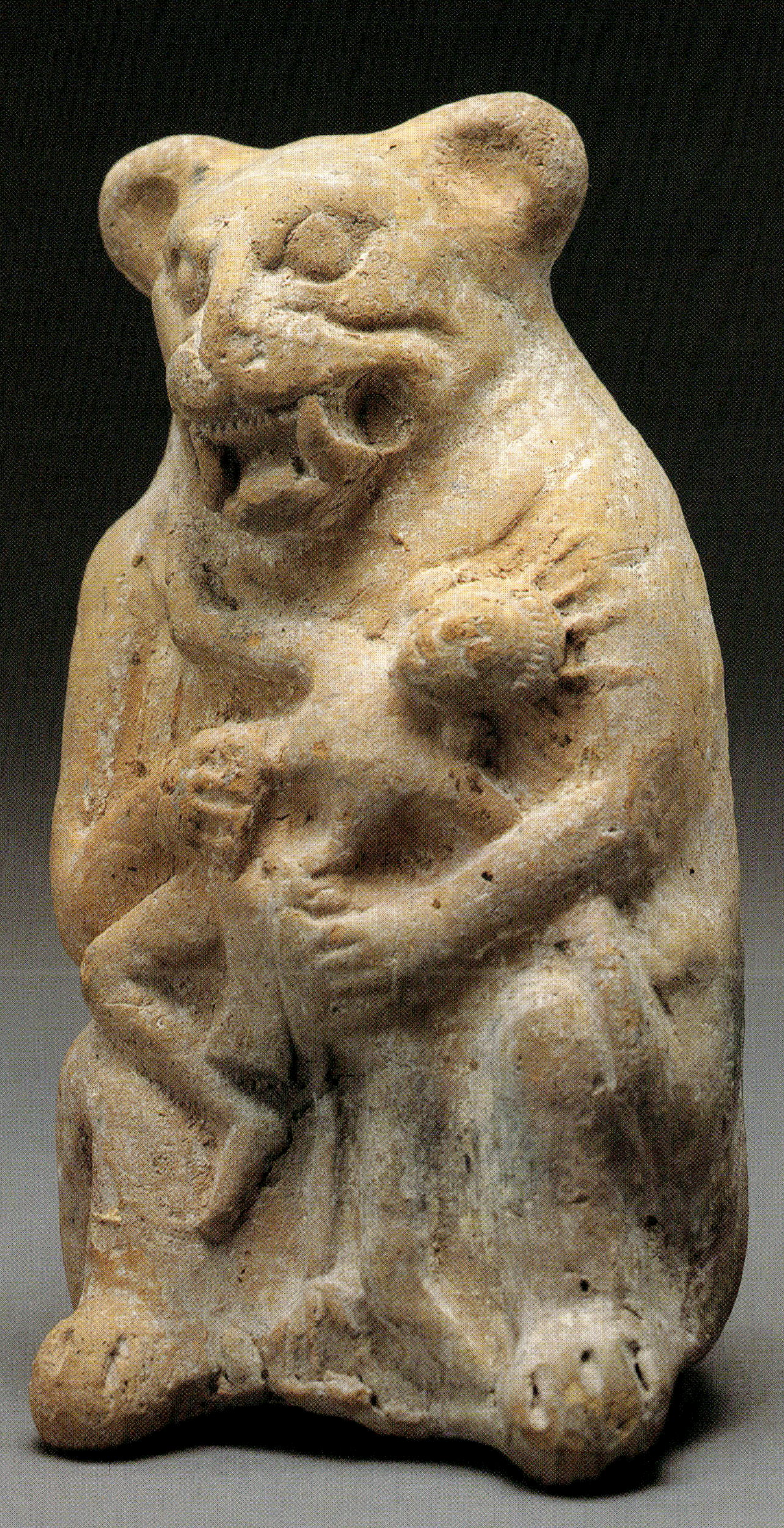

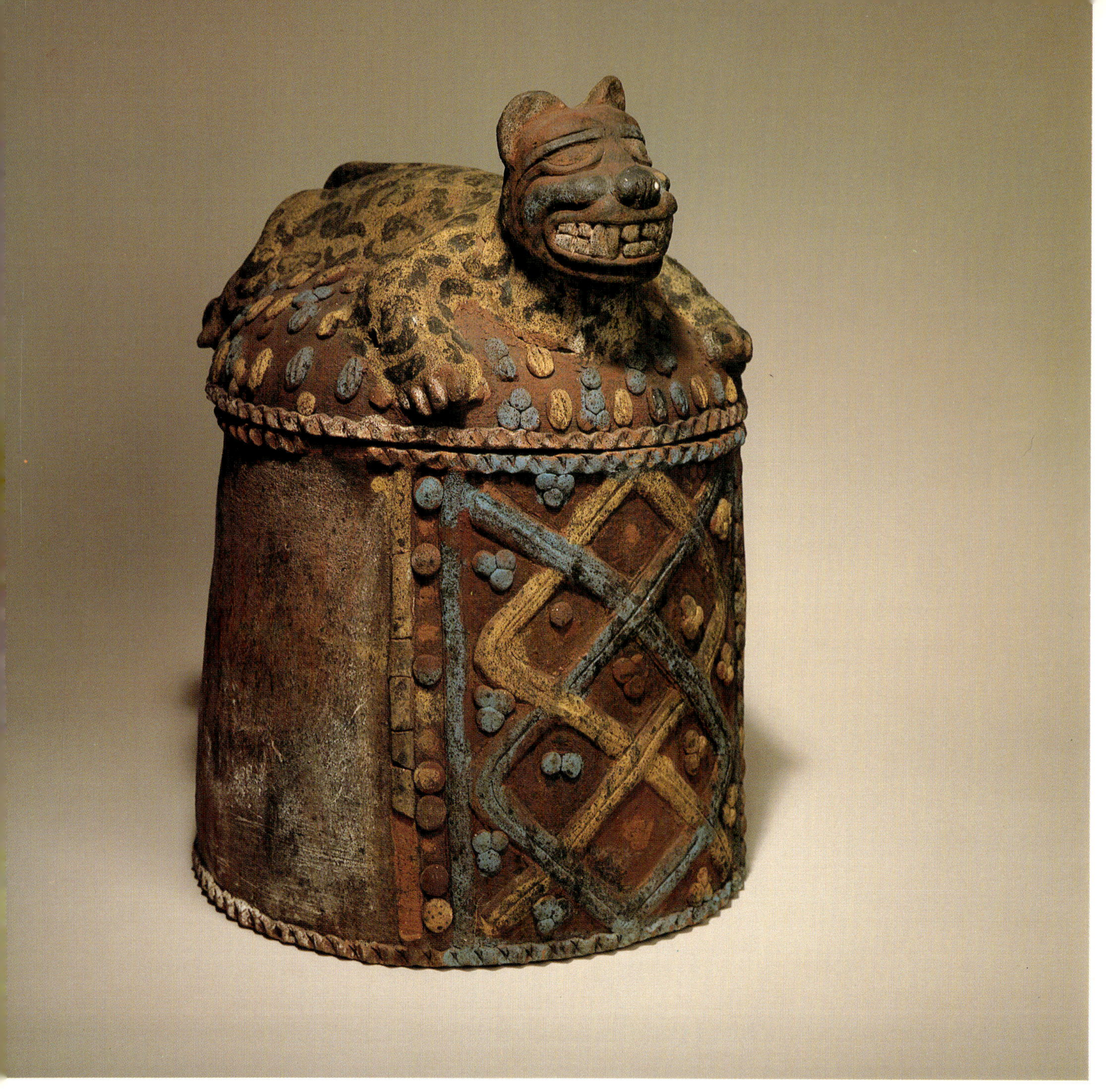

URN WITH JAGUAR LID   70

Maya. Nebaj, Quiché region, Guatemala Highlands
Late Classic: c. A.D. 600–900
Buff clay with yellow, blue, black and red paint. H:
15½″
The Bowers Museum, Santa Ana 78.75.14 a & b

On the vessel are modeled and painted crossed bands
of the mat glyph (pop) of rulership which when combined with the jaguar, was a maya symbol of royalty. (In Von
Winning & Stendahl 1968: Fig. 427; Labbé 1982: cat.
#97).

FIGURE IN JAGUAR COSTUME   69

Maya. Quiché region, Guatemala Highlands
Late Classic: c. A.D. 600–900
Buff clay with red, white, and blue paint. H: 6¾″
Collection of The Denver Art Museum, gift of Mr. and
Mrs. Edward M. Struss 1984.521

There is strong resemblance between this elaborately
modeled Maya figure and one of the Hero Twins in the
Popol Vuh, Xbalanque (Little Jaguar Sun). Both
Xbalanque and his twin, Hunahpu (One Blowgunner)
frequently went hunting with their blowguns. This Maya
figure holds a blowgun in one hand and two birds in the
other. Further, his face peers from the open jaws of a
jaguar headdress and jaguar claws dangle from his wrists
and ankles, just as Xbalanque is often identified by his
jaguar markings.

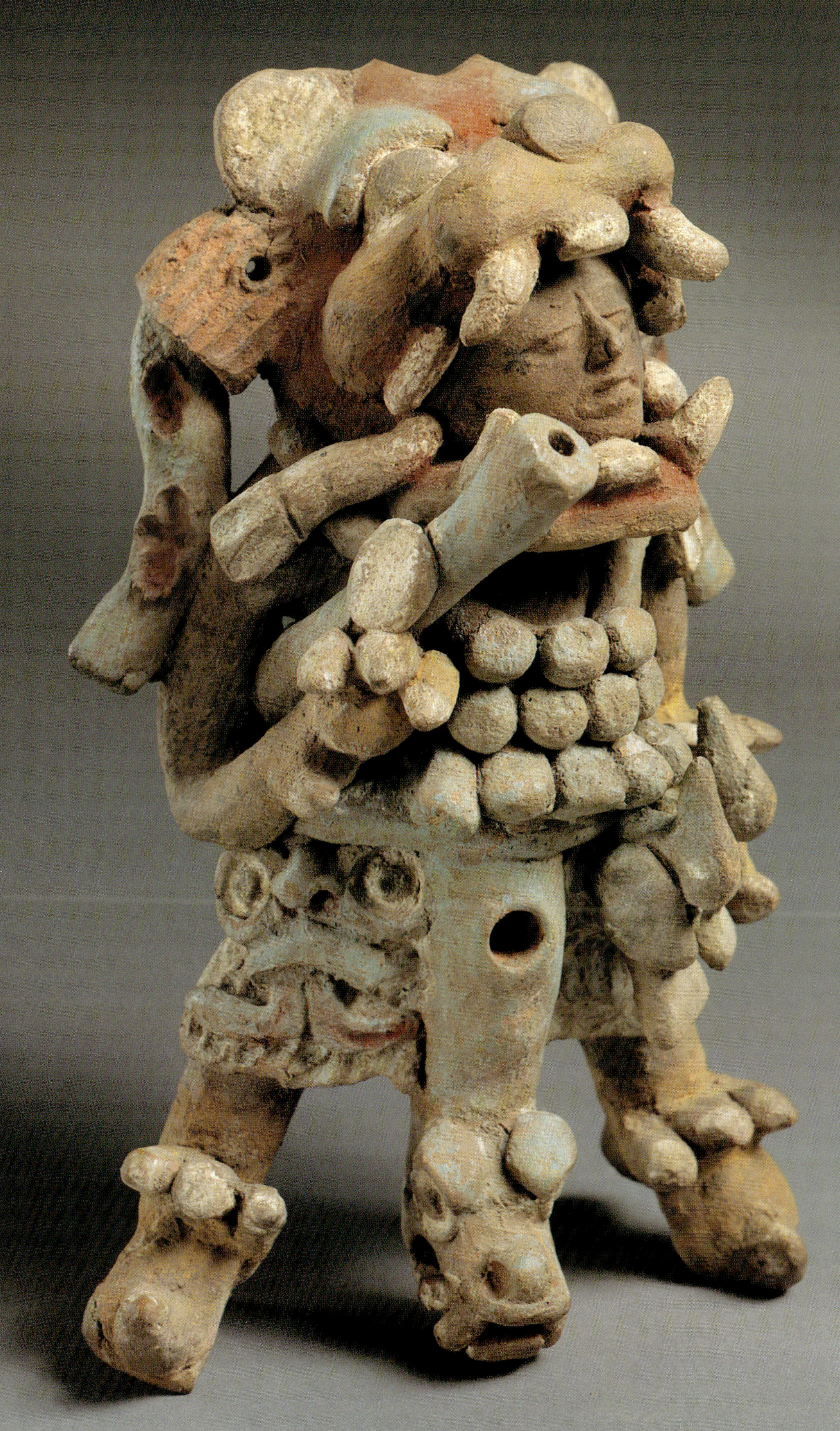

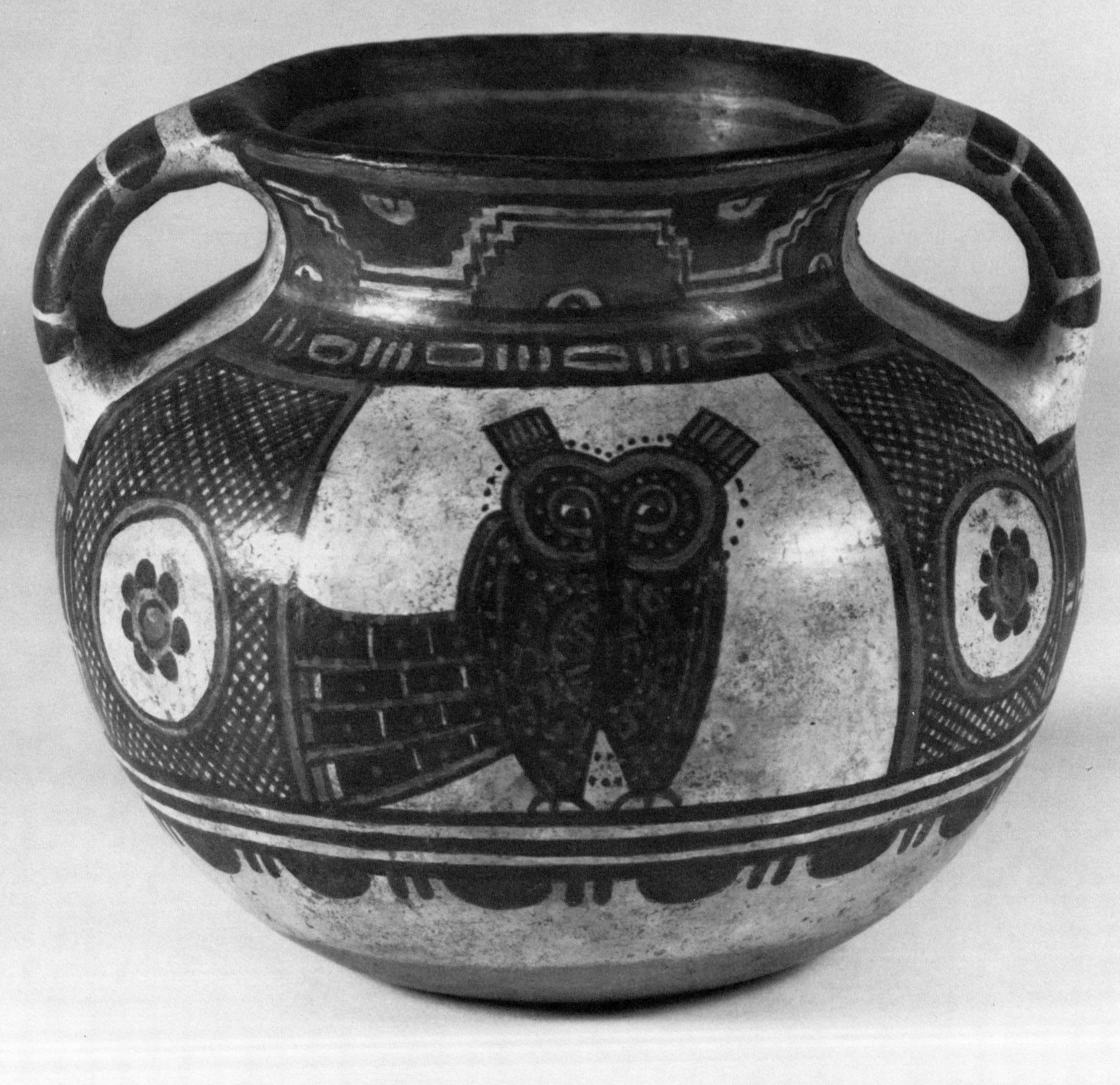

POLYCHROME VESSEL WITH OWLS   72
Maya. Ulua Valley, Honduras
Late Classic: c. A.D. 600–900
Orange clay with cream slip and black, brown,
orange paint. H: 6″
Collection of The Denver Art Museum, gift of
Frederick R. Mayer 1973.221

"MUSHROOM STONE" WITH JAGUAR   71
Maya. Guatemala Highlands
Late Preclassic: 100 B.C.–A.D. 200
Andesite. H: 12″
The Land Collection PL 408
(In Nicholson & Cordy-Collins 1979:155)

Several theories on the meaning and purpose of these
stones include their depiction of hallucinogenic
mushrooms. Mushroom stones may have also functioned as molds for clay bowls or rubber balls.

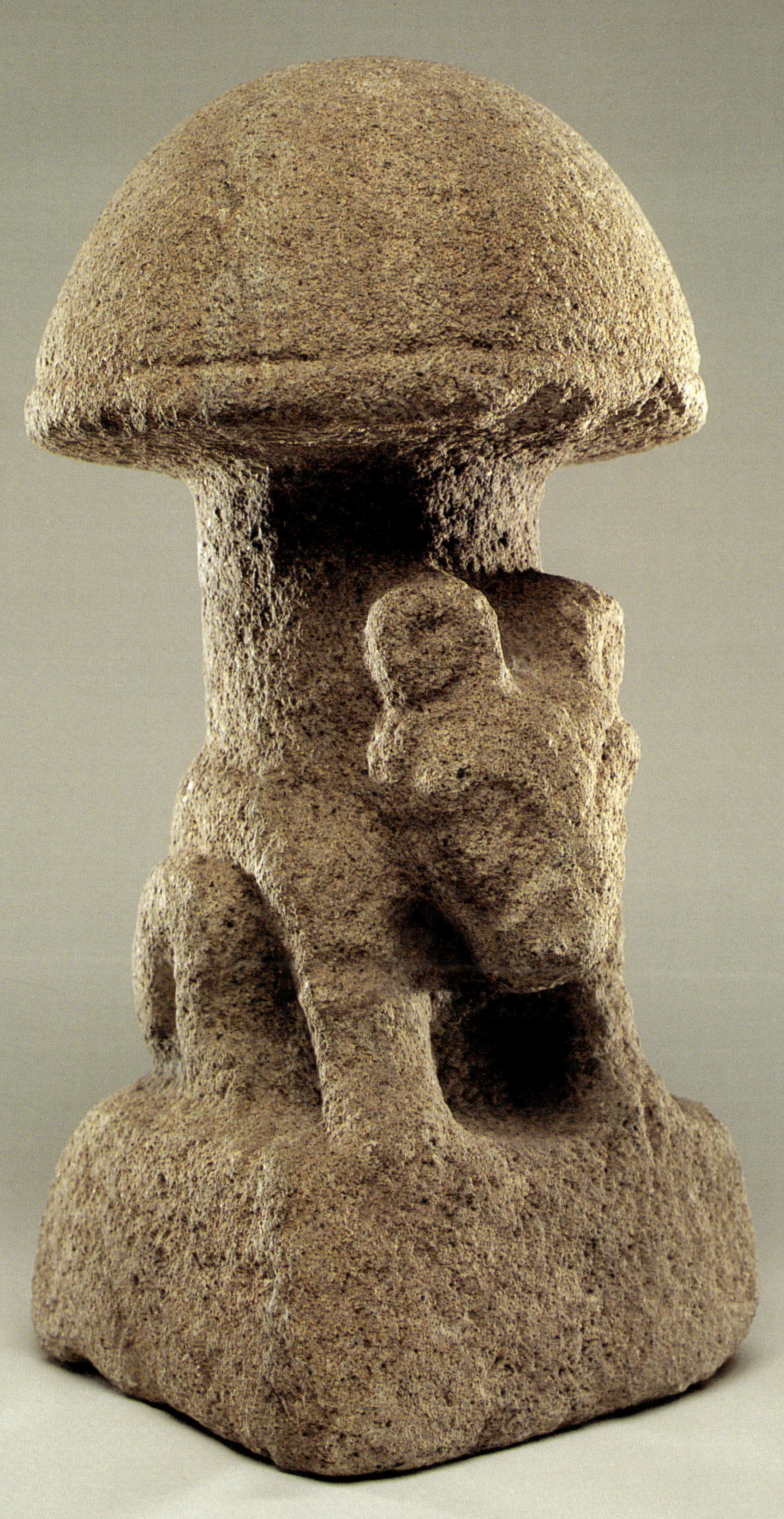

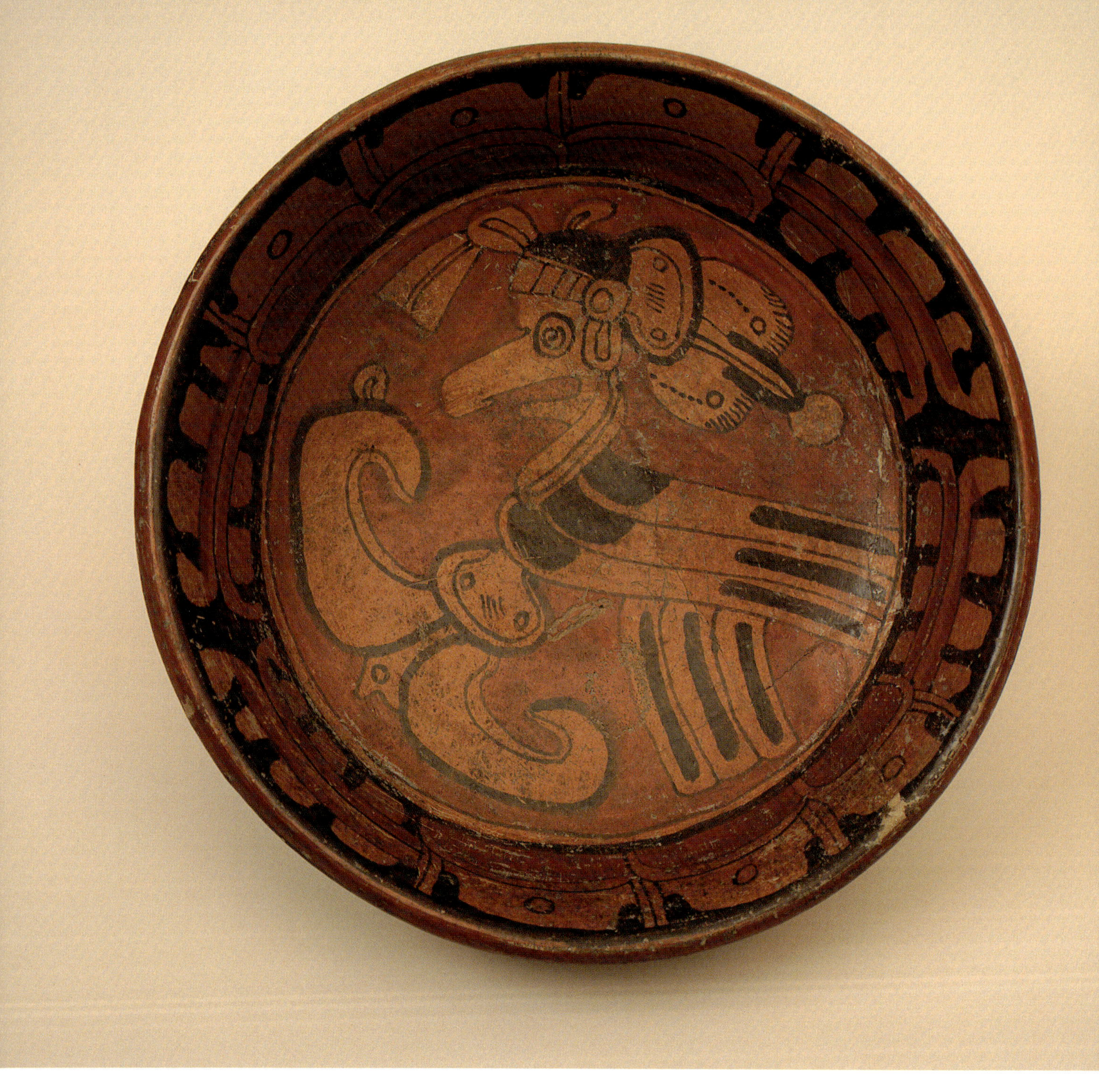

OWL-LIKE BIRD PLATE 73

Maya. Campeche or Yucatan, Mexico
Late Classic: c. A.D. 600–800
Cui-orange polychrome with black paint. D: 11½″
Stendahl Galleries

An owl-like bird is the supernatural featured on a large corpus of Late Classic earthenware plates from Campeche. This ware is named Cui-orange polychrome after the Yucatec word for screech owl or cui. Owl features of the cui include the rapacious hooked beak, circular eyes, and black-spotted feathering, similar to the Moan bird's. As an ominous sign of death and disaster, the nocturnal owl is perfectly suited to represent the underworld.

SEATED BAT GOD 74

Zapotec. Oaxaca, Mexico
Mid Classic (Monte Alban III): A.D. 300–650.
Buff clay. H: 11″
Linda and Steve Nelson

The Zapotec bat god was represented without wings but with human limbs. Its head has the exaggerated projection of the ''leaf nose'' bat above the nostrils, and an open mouth reveals the pointed canines of the vampire. This bat god holds a knife that may be sacrificial. The clay sculpture was uncovered in a tomb paired with the goddess J, who wore the same collar of pierced oliva shells.

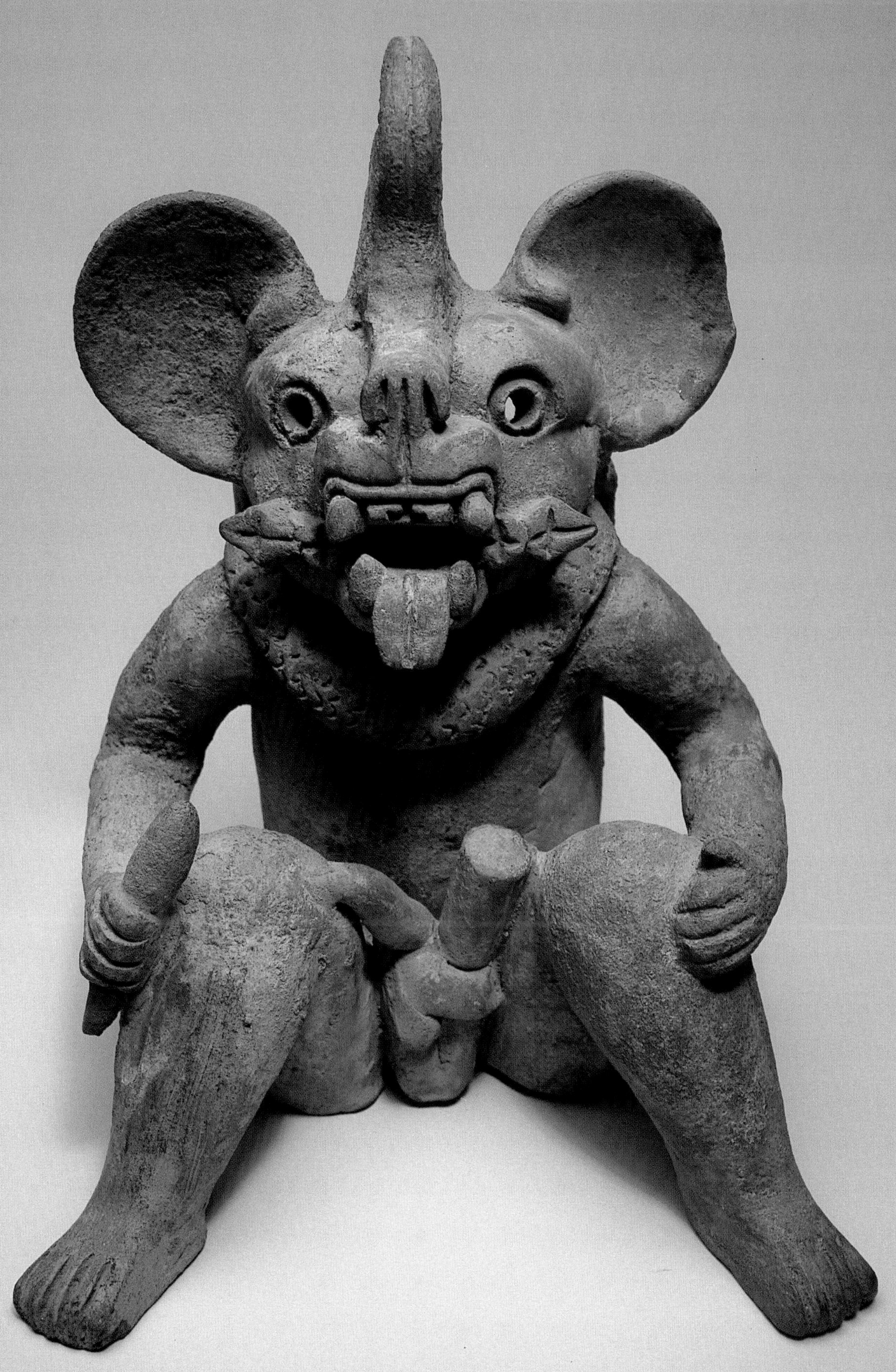

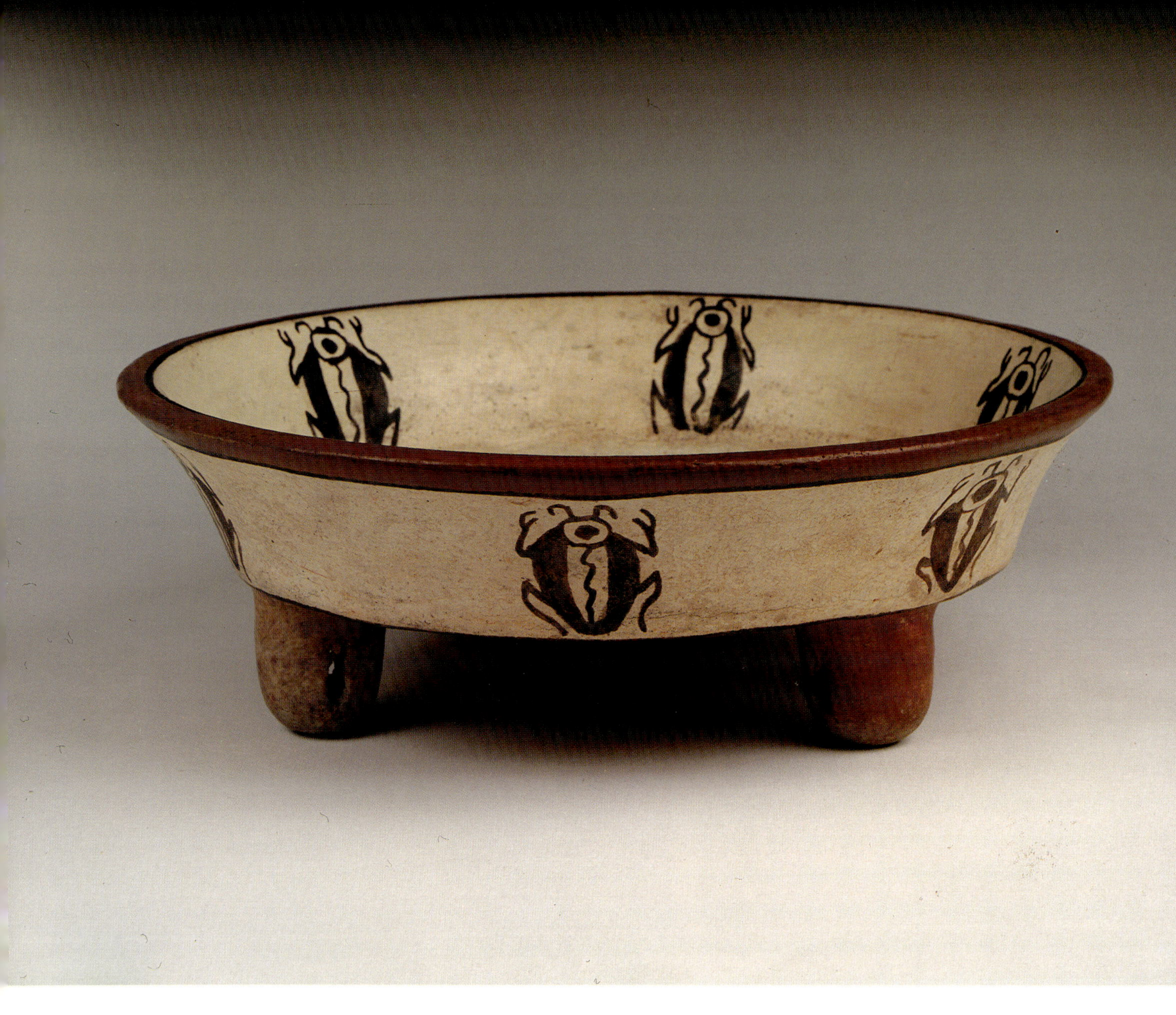

VESSEL WITH BLACK BEETLES   76

Maya. Tikal (Mundo Perdido), Petén, Guatemala
Late Classic: c. A.D. 600–800
Polychrome terracotta, black and red on cream.
H: 3½″
Museo Nacional de Arqueología y Etnología,
Guatemala 11141

PECCARY PLATE   75

Maya. Guatemala lowlands?
Orangeware with black and dark red slip design.
D: 11″
Early Classic: c. A.D. 300–400
San Diego Museum of Man, Wray 16–65

As a symbol of the underworld, the peccary or wild pig was a fitting image to be buried with the dead. The so-called ''kill hole'' in the center of the clay plate was intentionally pierced at the time of burial, perhaps to release the potent spirit of the plate itself and its imagery. Since clay vessels and plates were often placed over the head of the deceased, the 'kill hole' may have also allowed the soul of the dead to rise upward out of the underworld.

The peccary is sometimes shown breathing fire or smoke and wearing a sacrificial scarf. The sacrificial associations may have related to the peccary's dangerous teeth that can inflict a deep wound even on a jaguar. The peccary was associated with the four-sided Maya cosmos. When stylized peccary heads form the legs of a tetrapod vessel, they appear as four pillars supporting the world or the heavens (Schele and Miller 1986:280).[25]

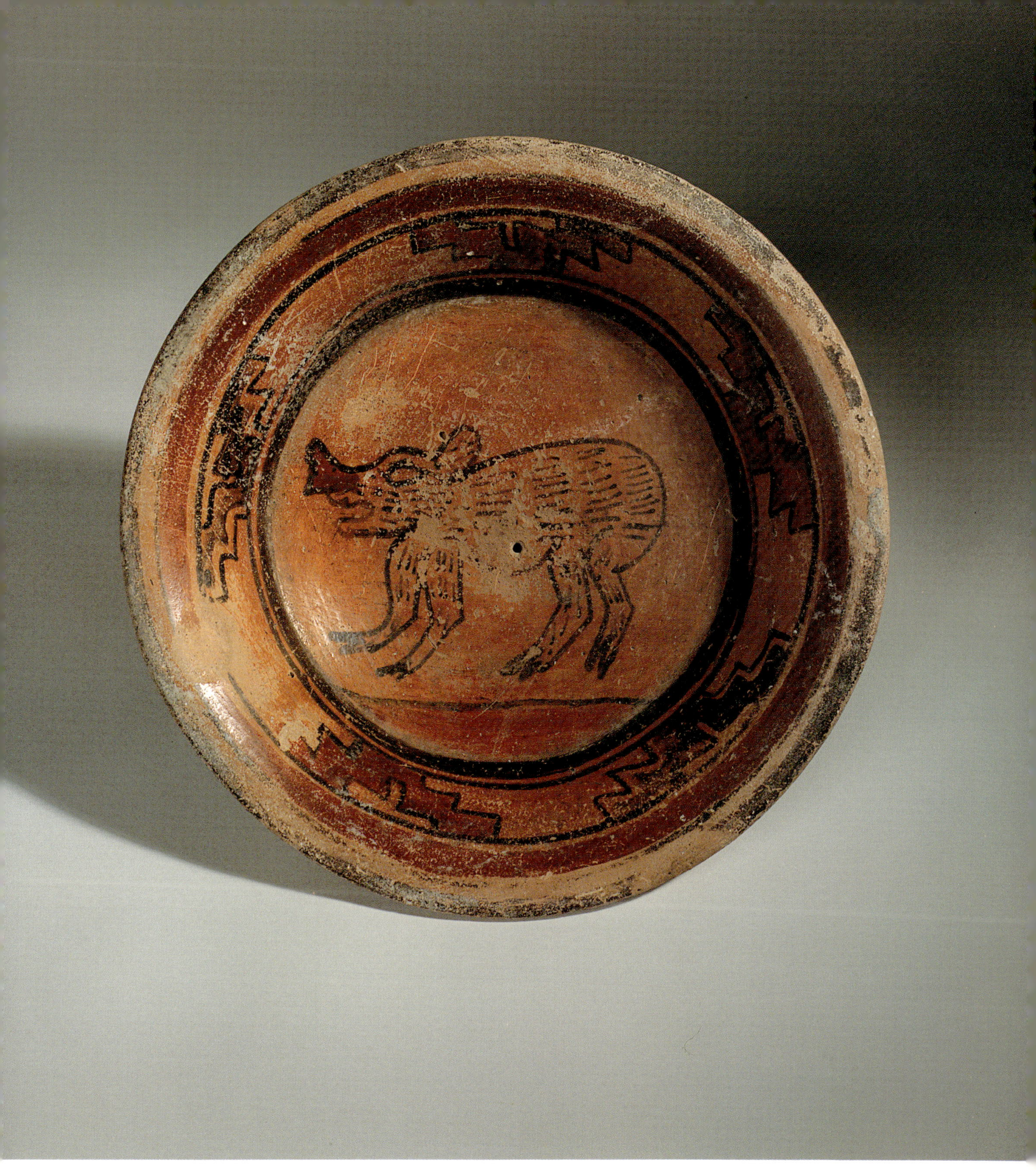

CONCLUSIONS

Plants and animals were selected as cultural and cosmic metaphors based primarily on their habitat and natural features. Yet, because of their numerous, overlapping meanings, they elude strict classification. The Mayanist, Eric Thompson (1970:197) acknowledged a similar difficulty in ordering Mayan religious ideas by noting that, "our containers for them are apt to turn into colanders."

The universe was viewed as an interrelated chain, dominated by the recurring cycles of life and death, day and night, rainy and dry seasons. Within this interactive universe, plants and animals assumed both creative and destructive meanings. The arboreal monkey was associated with the vitalizing sun and the arts, yet, as mankind's ancestor, also had connotations of death and the underworld.

Those creatures who in themselves embodied contradictions were put in a special category. The anomalous bat is a mammal that reverses the "natural" order by not only flying, but by hanging upside-down, thus peering into the underworld from the sky. As a nocturnal pollinator, the bat is the day hummingbird's counterpart (Benson 1988a: 112); both were associated with fertility and sacrificial death. Composite beasts who displayed bird, serpent, and jaguar elements, symbolized the transcendent natural forces of different cosmic zones.

Precolumbian costume elements (headdress, mask, and dress) played an important role in communicating these associations. The elite borrowed the plant and animal symbols that affirmed their ties with cosmological powers, especially with the sky and underworld (Fig. 56). When priests or kings donned ritual dress they were not only protected by these forces but were thought to actually become that god or supernatural (Fig. 33). Zoomorphic and botanic elements in a wearer's dress might also affirm his or her name, lineage, professional status and social rank.

Although plant and animal sculptures are intrinsically appealing and establish common bonds of familiarity, their meaning is often less accessible. Tragically, many Precolumbian art works are looted and have no provenance. Without the benefit of an archeological context, we will always be handicapped in our understanding of the work's original meaning and function. Works uncovered in burials, for example, may relay a culture's concept of the afterlife or reveal the identity and earthly activities of the deceased. When found in or near architecture, sculpture can be linked to the builder's ancestry, political aspirations, and religious beliefs. However enigmatic these art works remain to the twentieth-century viewer, we can be certain that all flora and fauna sculpture was charged with rich associations for its Precolumbian audience.

Man has always been dependent on his environment for his survival. More than ever, the environment is now dependent on man for its survival. Preserving the native American past is equally as pressing. In some small way, this exhibiton and publication are intended to help perpetuate our knowledge of Precolumbian Mesoamerica and remind us of the interdependence of man and nature, in order to promote the respect and survival of both.

ENDNOTES:

1. On **Metaphor:** Urton 1985; Isbell 1985; Heyden 1986.
2. On the **Precolumbian Cosmos:** León–Portilla 1963: 25–61, 1973: 56–90 and 113–159; Nicholson 1971a: 403–408; Klein 1976: 29–38; Clemency Coggins, in Clancy et al. 1985: 48–55.
3. On **World Tree:** Tozzer 1941: 43, 131–132, 136, 139; León–Portilla 1973: 73–77, 136–139; Shafer, n.d.; Schele and M. Miller 1986: 182, 268–269, 277–278; Coe 1988.
4. On Tree or **Pole Ceremony:** Redfield 1936; Sahagún 1950–1982, III: 104–107; Durán 1971: 203–207.
5. On **Vulture:** Seler 1909, IV: 49–52; Thompson 1970: 235, 337, 366–367; Blaffer 1972: 79–85; Aguilera 1985: 62–66.
6. On **Hummingbird:** Thompson 1970: 364–371; Hunt 1977; Benson 1988a.
7. On **Parrot:** Furst 1978: 27; Gallagher 1983: 35, 109.
8. On **Deer:** Lumholtz 1902, II: 45, 133; Seler 1909, IV: 538–541; Thompson 1970: 363–370; Klein 1976: 173–174; Aguilera 1985: 25–27.
9. On **Principal Bird Deity:** Bardawil 1976; Schele and M. Miller 1986: 55; Hellmuth 1987: 364–366; As "Seven Macaw" see D. Tedlock 1985: 86, 92–93, 360; M. Coe in Kerr 1989: 162–164.
10. On **Two–Headed Serpent:** Lumholtz 1902, II: 234–235; Pasztory 1983: 162, 172–173; Schele and M. Miller 1986: 106, 111; Carlson 1988a.
11. On **Serpent:** Lumholtz 1902 I: 310, II: 161–165, 191, 214, 234–235; Thompson 1970: 178, 252–258, 262–269, 274; Sahagún 1950–1982, IV: 50, 59; XI: 79, 81; Pasztory 1983: 233–234; Aguilera 1985: 72–75; Schele and M. Miller 1986: 72–73.
12. On **Frog, Toad:** Thompson 1970: 120, 166, 258–259, 268; Nicholson 1971: 406; Sahagún 1950–1982 II: 62; Furst 1981; Kennedy 1982; Joralemon 1988: 48.
13. On **Monkey:** Seler 1909 IV: 456–459; Thompson 1970: 237, 301, 322, 331–348; Sahagún 1950–1982 IV: 73–74, XI: 14; Pasztory 1983: 216, 223; Aguilera 1985: 40–42; D. Tedlock 1985: 83–86, 105–124.
14. On **Armadillo:** Lumholtz 1902 II: 165; Real n.d.; Carlson 1988b: 66.
15. On **Gopher:** Thompson 1970: 324, 336; Aguilera 1985: 30.
16. On **Rubber Ballgame:** Leyenaar 1978; Leyenaar and Parsons 1988; Schele and M. Miller 1986: 241–264.
17. On **Dog:** Sahagún 1950–1982 III: 40–42; Thompson 1970: 137, 188, 250, 300–301; Nicholson 1971a 413–414; Gallagher 1983: 35; Aguilera 1985: 20–23.
18. On **Maya Underworld and Waterlily:** Puleston 1977; Schele and Miller 1986: 135–138, 141–142, 265–288, 302; Hellmuth 1987.
19. On **Waterbird and Duck:** Furst 1978: 27; Gallagher 1983: 35; Schele and M. Miller 1986: 55, 68; Hellmuth 1987: 352.
20. On **Conch Shell:** León–Portilla 1963: 107–109; Furst 1978: 13–15; Schele and M. Miller 1986: 54, 83–84; Coe 1988: 231.
21. On **Jaguar:** Sahagún 1950–1982, XI: 1–3; Furst 1968; Thompson 1970: 220, 234, 291–294; Coe 1972; Pasztory 1983: 172, 207–208, 233; Benson 1985, 1988b, In Press.
22. On **Owl:** Klein 1976: 126–130; D. Tedlock 1985: 272, 347; Clancy et al. 1985: 180–181.
23. On **Bat:** Caso and Bernal 1952: 67–100; Blaffer 1972; Aguilera 1985; Benson 1988a.
24. On **Tapir:** Tozzer 1941: 203; Thompson 1970: 335, 354; Aguilera 1985: 27.
25. On **Peccary:** Robicsek and Hales 1981: 47–48; D. Tedlock 1985: 339–340, 369; Clancy et al. 1985: 170, Pl. 114; Schele and M. Miller 1986: 55, 280.

SCORPION EFFIGY   77
Colima, West Mexico
Late Preclassic–Early Classic: 200 B.C.–A.D. 300
Monochrome red. L: 7⅞″
The Land Collection ML 347
(In Dwyer & Dwyer 1975: Fig. 64)

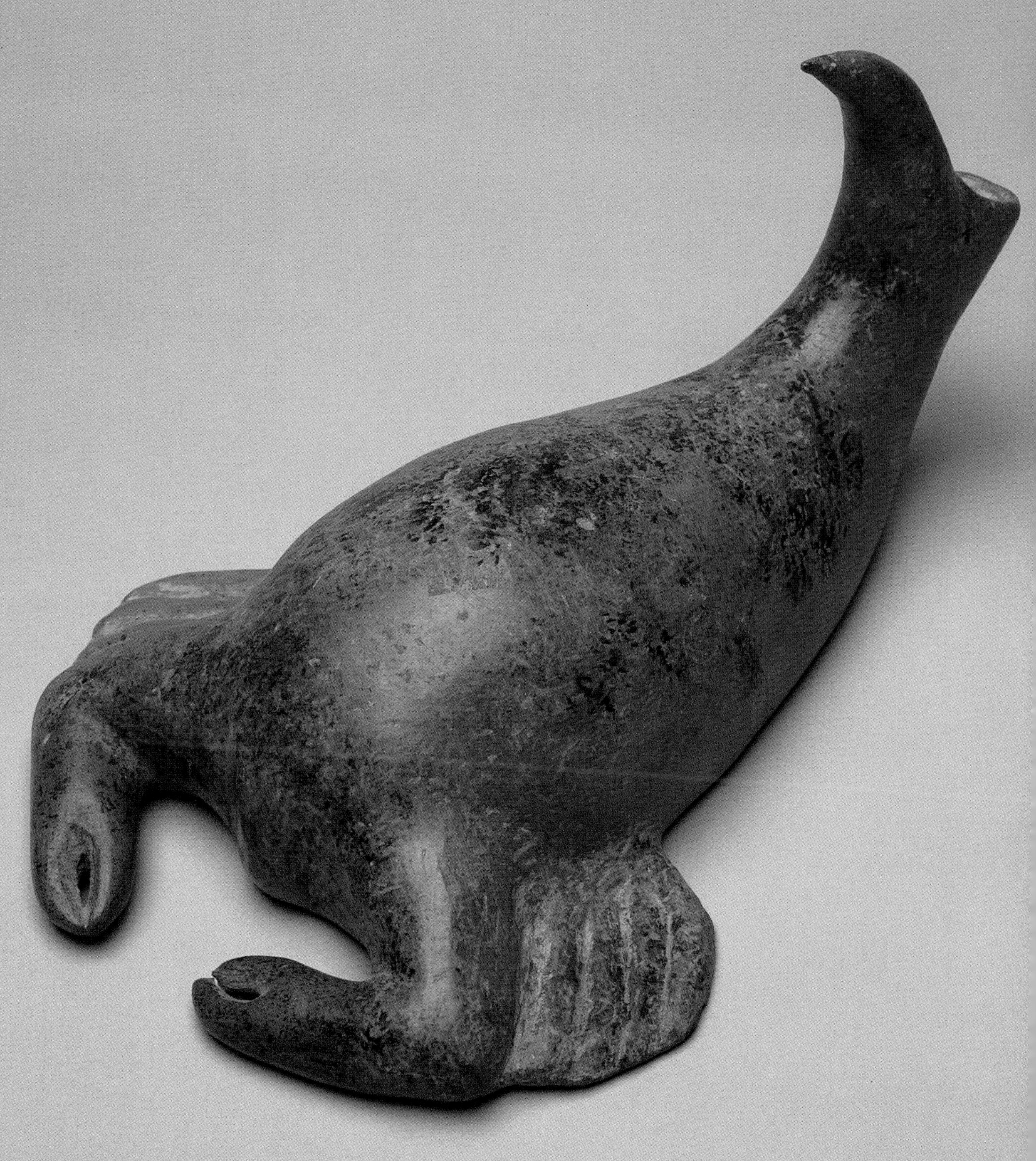

ANIMAL AND PLANT THEMES IN MESOAMERICAN RITUAL AND FOLK ART

Introduction

There is a timeless and personal quality about animal and plant motifs in art, and Mesoamerican folk art is no exception. They reflect man's close association with the natural world — the living environment from which he takes sustenance and comfort. Sometimes animals and plants serve as symbols in a culture's belief system, or world view. Parallels have been discovered between the recorded beliefs, myths, ritual and art objects of ancient Mesoamerica and those of contemporary Indian societies. Examples abound of the use of flora and fauna in Mexican and Central American ritual and folk art, such as the Flying Pole Dance, the masked jaguar and the cosmic trees which are direct and proven survivals of ancient ritual and myth, although the original religious significance may be modified or lost.[1]

It has been over 400 years since the Spanish baptised thousands of Indians into the Catholic faith in Mexico and Guatemala shortly after the Conquest (1521). Yet, the roots of the Preconquest Indian belief system were deeply entrenched and not completely eradicated. Anthropologists have uncovered many survivals of the ancient native world view and religious themes in areas of the same regions where ancient civilizations once flourished.

Other examples of Indian Preconquest plant and animal themes have been intertwined so firmly with Spanish Catholicism, that the native religious symbols and the introduced Christian ones are distinguishable only with close study. In a process called syncretism, certain elements of an introduced culture became incorporated into an existing one to produce a combination of both, as manifested in a ritual, dance or even a single object. This may not result in an easy "blend" and it does not necessarily reflect a stable or static situation culturally (Tedlock 1983:243).

Other plant and animal themes cannot be identified, at present, with ancient symbols. They may reflect solely the inspiration of a particular folk artist by the living creatures of his environment. A brief introduction to the artist and techniques of his or her art is followed by the three categories of animal and plant themes:

1) Those which are mainly survivals of native (Precolumbian) beliefs, the primary emphasis of this essay,

2) Those which are syncretic (combined European and Indian) elements, and

3) Themes relating to animals and plants with no detectable Prehispanic elements. These comprise a large body of folk art in which the artist has used flora and fauna themes, but the origin of these themes is not known at present. Was the artist inspired to make this object simply by his or her surroundings, or is there a deeper meaning?

The Folk Artist and Folk Art of Mesoamerica

Folk art is the art of the people. In Altman's (1961:356) definition he characterizes it as the art form developed by those outside of the ruling elements of Asiatic or European society. Folk art is in definite contrast to the art production fashioned for an upper class in an hierarchical society. Tourist inspired productions belong in a category of their own, and thematically do not fit into this context.

The folk artist in Mesoamerica is a rural dweller or has rural roots. He or she may or may not speak an indigenous language or be identified as an Indian, but exposure to Indian culture could, no doubt, be established. The artist's life is closely knit with the local flora and fauna. Most villagers have a cornfield *(milpa),* and roaming goats and chickens are the same ones that produce the milk or meat for the table. Wild game hunting remains an important food source.

Pottery making, handweaving, metalwork, basketry, woodcarving, lacquerwork, papermaking, floral decorating and many minor arts still flourish today in Mesoamerica today using ancient techniques and materials. Pottery making, for instance, includes the Prehispanic techniques of handmodeling, coiling and molding clay The true potter's wheel was unknown in Mesoamerica before the Conquest. The pottery maker used slips and burnished his or her wares, but glazes were introduced by the Spanish. Weaving was done on a backstrap loom.[2] Featherwork, so important to the nobility in Preconquest times for the decoration of fine clothing, military banners and religious paraphernalia, is rarely found today.[3]

Basketry and mat *(petate)* making was widespread and the techniques used were primarily twilling, twining and tight coiling techniques (Foster 1967:123). Metalwork included the methods of hammering, repousee and filigree, as well as those of fusing, soldering and gilding metals. The metalworkers of Mesoamerica knew gold and silver, copper, tin, lead and even bronze. Indian artisans fashioned vessels of lacquered gourds, as well, by applying a tough, brilliant and moisture resistant coating. To accomplish this, they used a variety of oily or resinous substances bound to earths and colorants. (Fig. 79).

HOUSE CROSS IN A CRUCIFORM DESIGN WITH BIRDS   78

Zapotec Indian, Santo Domingo Xagaacia, Villa Alta, Oaxaca
Unfired clay. H: 19¾"
Collected in 1964 by Anita Jones
San Diego Museum of Man 1964-102-75

The crosses are placed on the roofs of homes for the protection of the occupants.

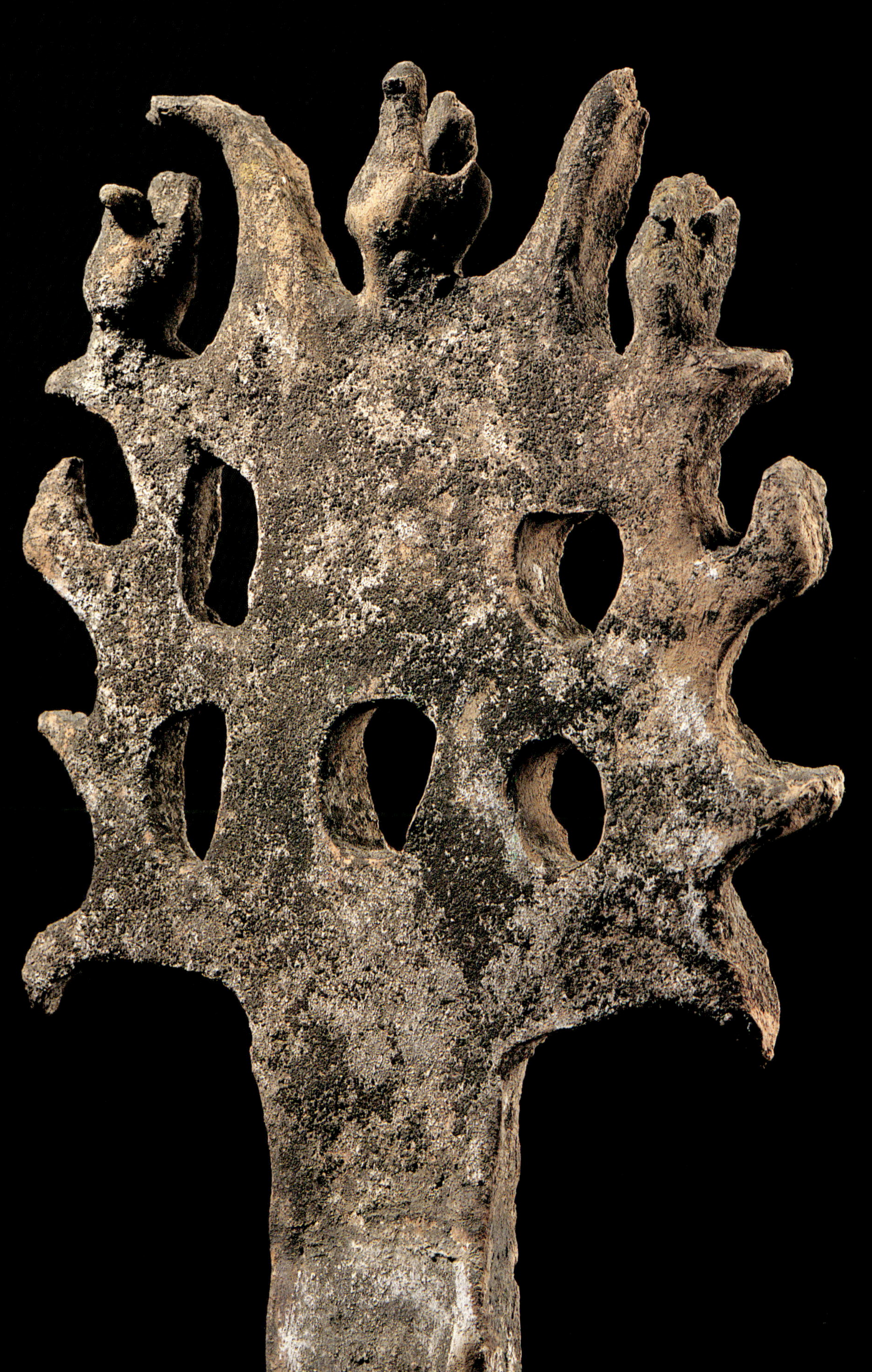

SURVIVAL OF NATIVE BELIEFS

Animistic nature worship was the foundation on which Mesoamercan religious beliefs were formed. Rituals performed today of this character (such as the deer dance of the Yaqui and Mayo Indians) are descendents of local native rituals allowed by the Spanish clergy through the ages because of the dance's "harmless" character. In some cases, they may be diluted versions of complex ceremonies presented originally by the priestly hierarchy. Eva Hunt (1977:272) describes the Indian religious systems as old ". . .versions of highly sophisticated structures, living and creative symbolisms that carry not just beauty but also fundamental functional visions of their life conditions and deep emotional commitments."

Key to our theme, is that human beings can influence the course of nature by propitiating the gods. This belief persists in many areas of Indian Mexico, probably because it seems to be effective in bringing rain, animals to hunt and good harvest. In Preconquest times, and today, this petitioning takes form in ritual. Thus it is in dance costumes, masks, and ritual paraphernalia that many identifiable survivals can be found.

Christian conversion of the Mexican indigenous population was outwardly more successful in some areas than others. The Indians were used to accepting new gods in addition to their own as they had under the Aztecs, so they readily accepted the new Christian god and saints to their pantheon of deities. But the Catholic Church rigorously punished the practice of worshipping multiple gods, so such devotions became furtive. Christ, the Virgin Mary and the Catholic saints took on the former attributes of the local Prehispanic gods. In many villages in Mexico and Guatemala the patron saint's image is treated as a deity who can cause good fortune or disaster to the people, much as the Precolumbian gods were perceived.

Another ancient belief was in the special relationship between man and animal. This is the understanding from birth that one has an animal "familiar," which is not a flesh and blood creature, but a spirit. The familiar is an alter ego of each person, often defined at birth and in some cases, not discovered by the human until adolescence after a type of vision quest. This alter ego is called a *tonal* (Adams and Rubel 1967:341). It is connected to the basic Precolumbian belief in an individual destiny which is fixed at birth. Certain powerful people may have a *nagual,* an animal into which they can transform themselves, sometimes for evil purposes. Belief in this concept is called *nagualismo.* Masks, costumes and ritual paraphernalia were used to effect that transformation for the dancers and his audience in a public way. The idea was, and is, to cause nature to do what man desires, using the power represented by the masked dancer and, in some cases, the spirit-animal imitated.

Ritual Survivals

Some areas of Mesoamerican popular art are without doubt survivals of the practices of ancient times. Why would certain traits survive over 400 years of exposure to European customs? In some regions, isolation was a factor. The Huichol of West Mexico and the Lacandon of Chiapas, for instance, were not strongly influenced by Christian missionization until the seventeenth century. Many areas of Northwest Mexico, Sierra de Puebla, Guerrero, Chiapas, Quintana Roo and parts of Guatemala are only sporadically visited by the clergy, even today. The local Indian groups in these areas still exercise considerable control over religious rites in their villages.

LACQUERED GOURD CONTAINER 79
Mexican, attributed to Indian artists from Michoacan, Mexico
Natural gourd with a silver hinge, snap lock and a wooden egg-shaped finial
Lacquered in the painted *(aplicada)* technique.
H: 16¼"
Probably second half of the nineteenth century
The Natural History Museum of Los Angeles County
L.2100.52-158

This delicately painted piece was probably used for religious purposes judging by motifs of a Bishop's mitre and crook, and on the reverse side an open book.

DEER HEAD FIGURE 80
Maya Indian, Western Highlands of Guatemala
Wood with metal bells and cotton. H: 22½"
Collected in 1975
Gordon Frost Folk Art Collection, Benicia. 8G41X

This figure was probably used in the Dance of the Deer or the Dance of the Tun according to the collector's notes. The Deer Dance is a ritual hunting dance performed in honor of the native gods of the Maya and dedicated to them.

DEER MASK 81
Maya Indian, San Marcos, Guatemala
Wood with glass eyes and deer antlers. 19" × 15½"
Collected in 1969
Gordon Frost Folk Art Collection, Benicia. 2G10

This mask was used in the Dance of the Deer. In this hunting dance-ritual which is considered to be of Precolumbian origin, the hunters simulate a deer hunt and with spiritual help capture their quarry. A similar dance is performed by the Mayo and Yaqui Indians in Northwestern Mexico.

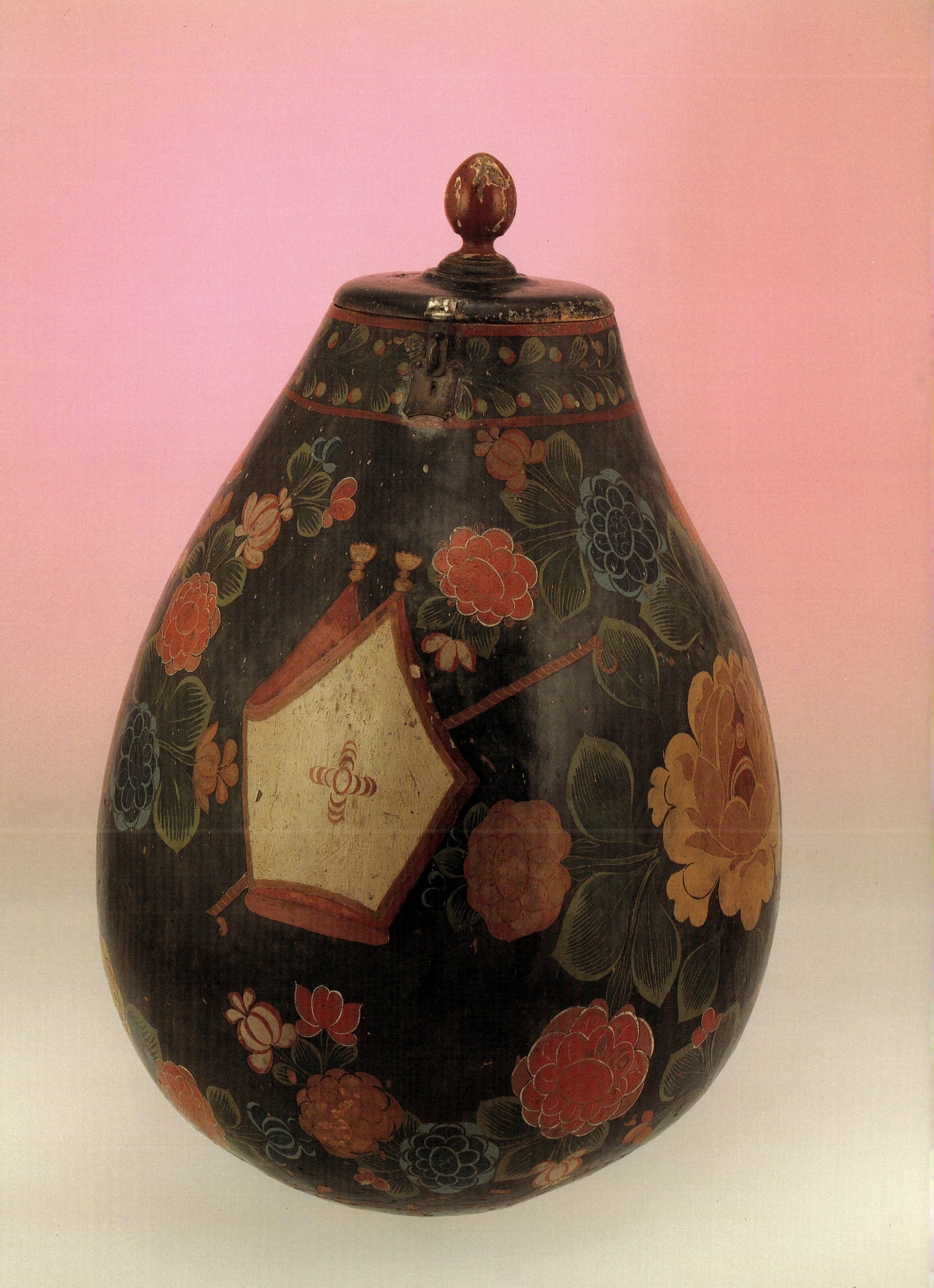

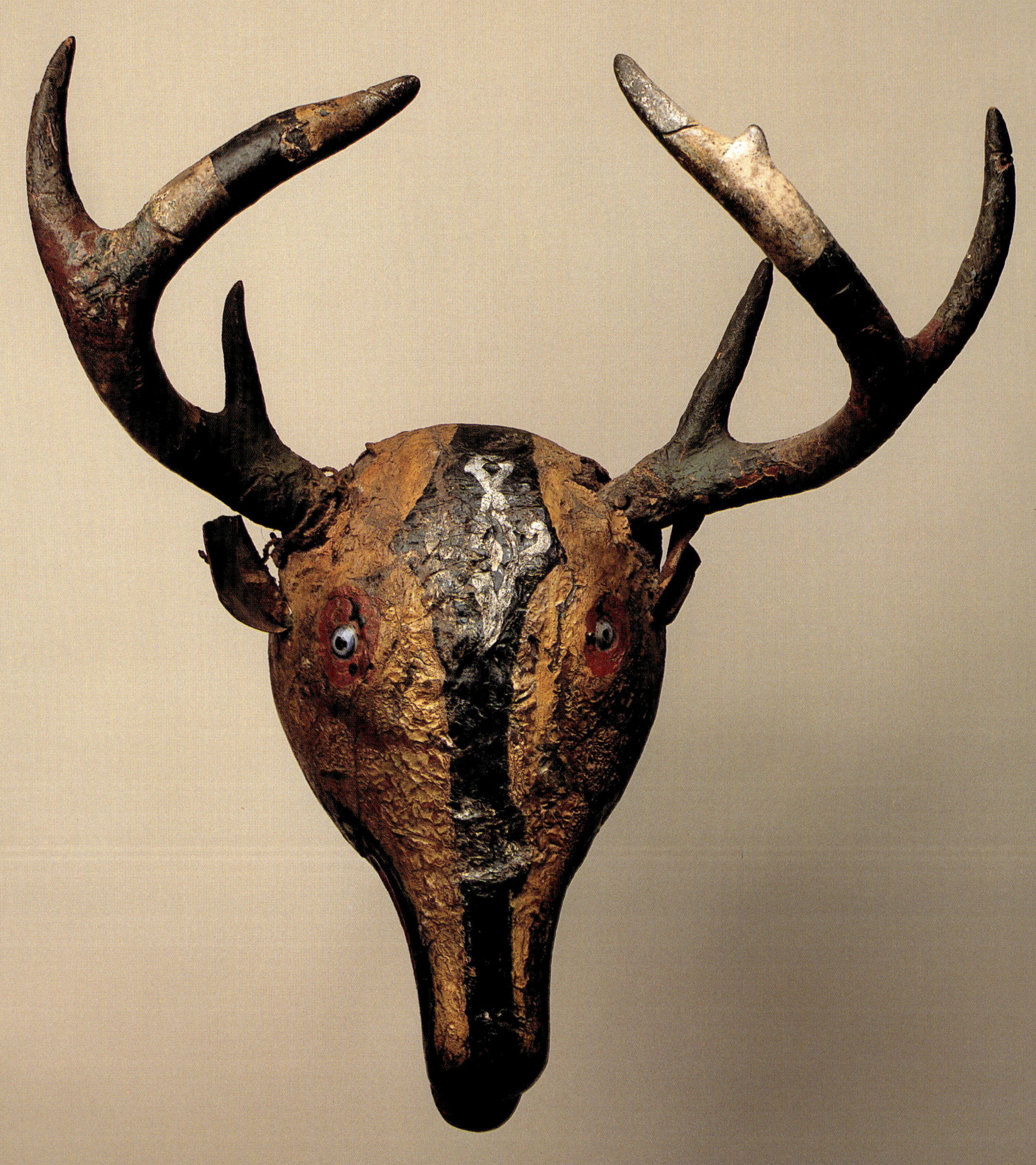

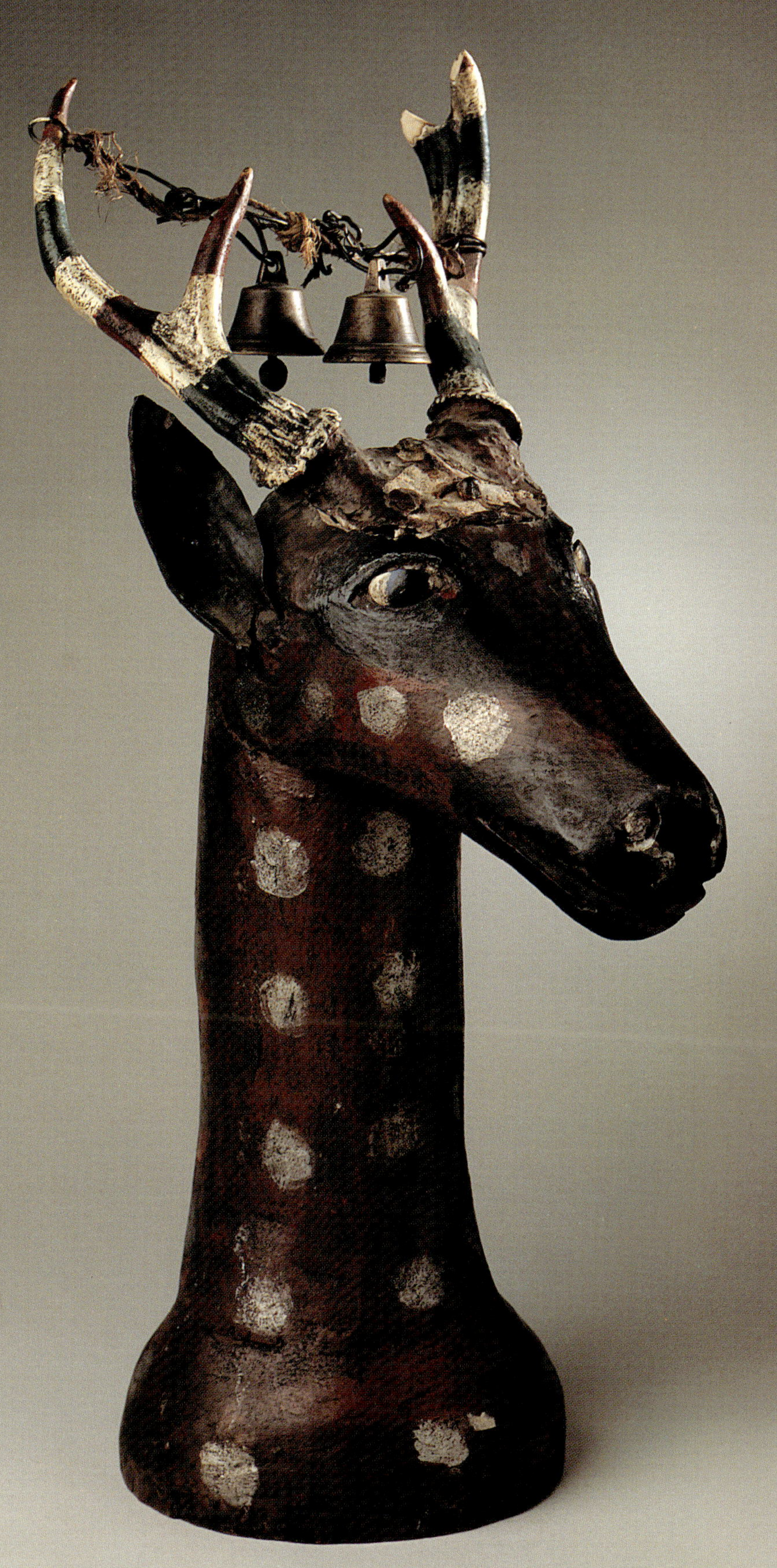

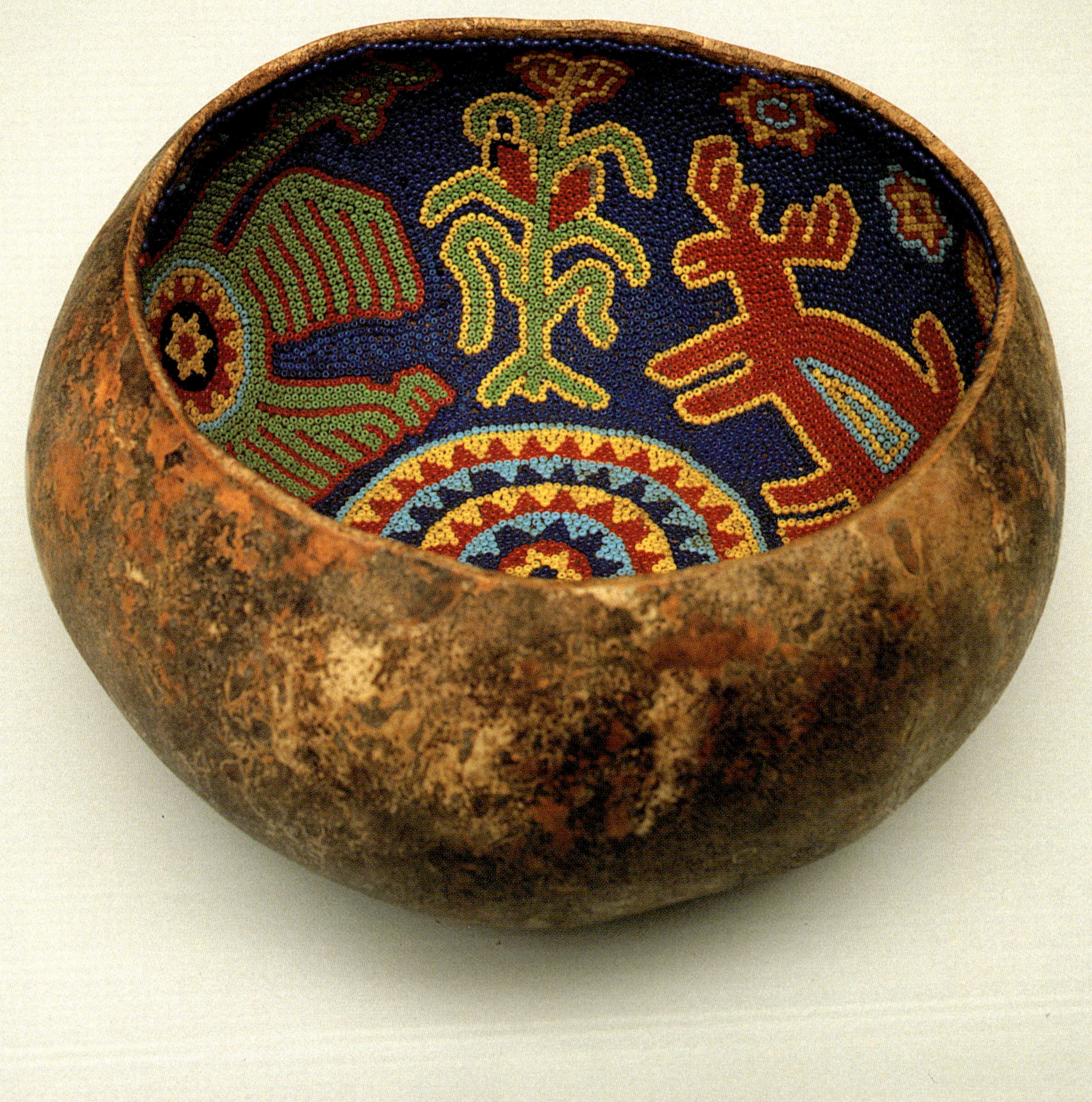

BOWL WITH CORN, DEER AND DOUBLE-HEADED EAGLE DESIGNS 83
Huichol Indian, San Andres Comiata, Jalisco, Mexico
Gourd decorated with glass beads adhered with beeswax. Dia: 6½″
Collected by Susan Eger Valadéz in 1983
San Diego Museum of Man 1983-44-2
Made by Pablo Carillo

MASK OF THE DOG KNOWN AS *LA MARAVILLA (THE MARVEL)* FOR THE DEER DANCE 82
Kekchi Maya, Western Highlands of Guatemala
8½″ x 4½″
Collected in 1972
Gordon Frost Folk Art Collection, Benicia. 5F260

The dog, companion of man, assists the deerslayers in the simulated hunt for the deer in this animal dance-ritual presumed to be of ancient origin.

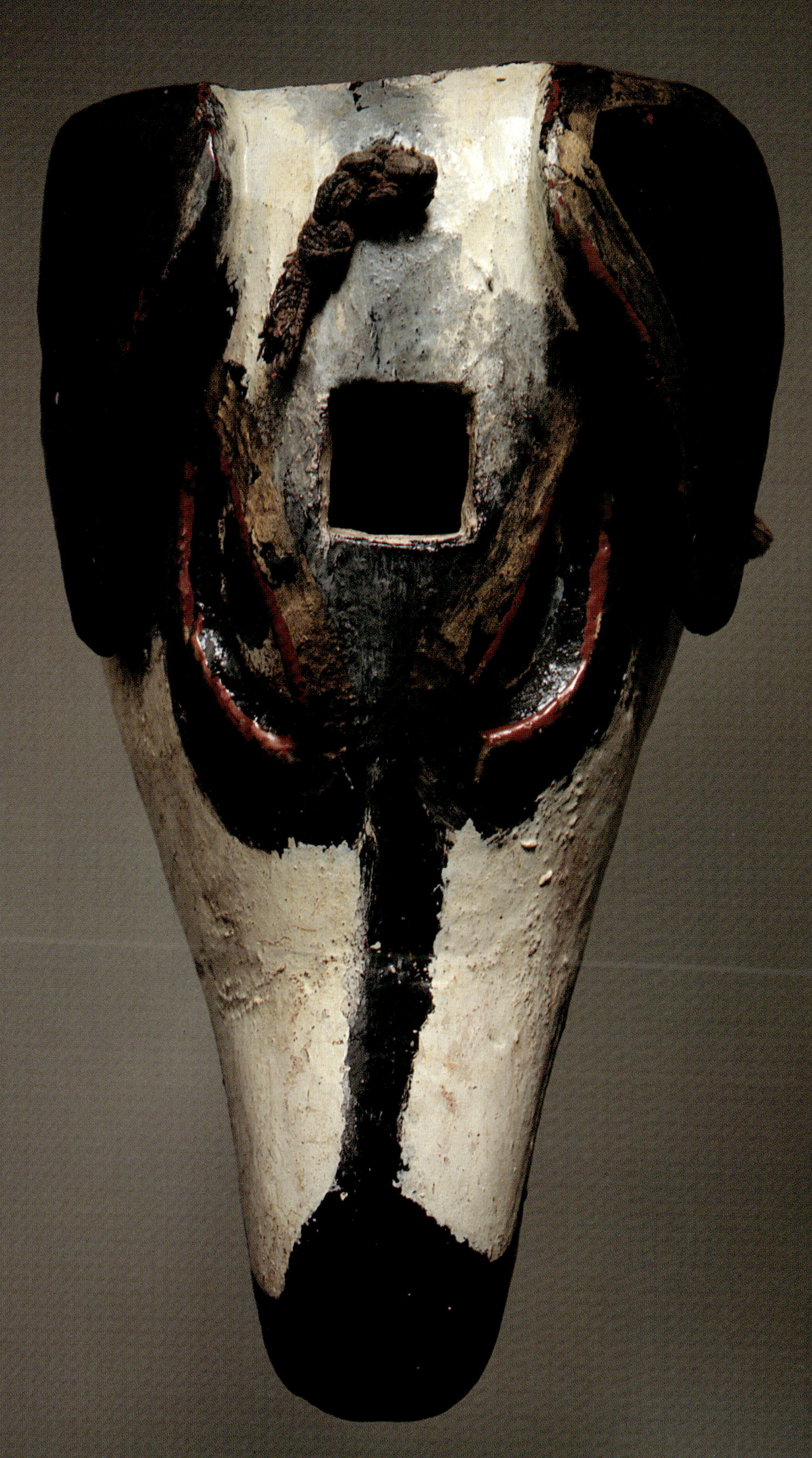

Another factor to account for survivals is "incomplete" conversion. The friars baptised thousands of Meso-american Indians in order to fulfill their mission and the Spanish Crown's orders, often not knowing the language of their converts. They used various strategies to encourage acceptance of Christianity, such as building Catholic churches on top of ancient sacred sites. The priests permitted natives to perform their dances in honor of the saints instead of the ancient gods, as long as they were not what they considered obscene or pagan. The Spanish priest Pedro de Gante wrote in the sixteenth century that he used the Indian custom of singing and dancing to their gods at sacred sites as a method of con-version. He composed songs with Christian themes to accompany their dances and provided the Indians with Christian designs to paint on their mantles instead of the Preconquest motifs honoring their deities (as cited in Madsen 1967: 376–377).

Perhaps the most significant factor leading to the survival of a ritual was the vital importance of their native religious belief to the people themselves. The Indians gave the old gods the names of Catholic saints, superimposing their traits and associations with natural phenomena. The old rituals were purged of cleric-offending parts and names or easily recognized symbols of the old gods. Even the ancient ritual calendar system is still in use among some of the Maya Indians of Chiapas and Guatemala. Sooth-sayers in Momostenango, Guatemala still use this system to determine the propitious times for fiestas and other special events.

Following are some examples of survivals in ritual along with the animals or plants associated with them.

The Deer Dance, the Deer and the Dog

The Deer Dance, La Danza del Venado, is performed in widely separated areas of Mesoamerica. The dance of the Yaqui and Mayo Indians of Northern Mexico is prob-ably the best known. Both that performed in Mexico and the one in the Western Highlands of Guatemala (Figs 80, 81) are hunting rituals that are widely accepted as Precolumbian in origin (Beals 1945: 202–204, Paret Limado 1963, Lothrop 1927, 1929).

Survival elements common to both include dedication of the Deer Dance to the native gods. Among the Quiché Maya the dedication is to the world god *Dios Mundo,* and among the Tzutujil Maya, the God of the Hill. (Paret Limado 1963:13). The hunters seek the shaman 's (spiritual practitioner's) help in the hunt of the deer, and request permission from the gods to obtain their quarry. Among the Yaqui and Mayo Indians, the deer dancer is bare-chested and clad in a white kilt with cocoon rattles on his legs, a leather belt with deer dew claws dangling and a genuine stuffed deer head as his headdress. He sniffs the air and paws the ground like a deer. He is track-ed and hunted with bows and arrows and eventually felled and "skinned" amid much joking and antics by other dancers. Ralph Beals (1945:204) surmises that mis-sionaries strongly encouraged this dance as "honest" (which meant no sexual elements or references to ancient gods). The Indians performed deer dances in churches and in some cases other villages were invited by the friars to observe performances.

The European clergy undoubtedly missed the significance of the animal theme and its connection to the Prehispanic deities. Beals (1945:190) reported in the 1930's that the Mayo and Yaqui Indians associated the Deer Dance with the "old religion," called the religion of the woods, rather than Catholicism. Perhaps the closest connection with ancient dance-dramas is the Aztec ritual conducted to honor the god Mixcoatl-Camaxtli (god of the hunt) where participants engaged in a ceremonial hunt of deer, coyotes and rabbits. Captors took the heads of their game home and hung them in their houses (Sahagun 1950–1981, Bk. II: 137).

In Guatemala, the Deer Dance includes Prehispanic elements of prayer to the four world directions, the use of native incense *(pom),* a simulated deer hunt, and in some cases the planting of trees. A tall tree was placed near the village church and cable was attached to it stret-ching to the ground at a 45 degree angle for characters in the dance drama to climb. In Texcaltitlán, Guerrero a jaguar dancer climbs the cable to the church tower in a desperate attempt to escape his pursuers (Markman and Markman 1989:170). Trees appear in several animal dances including that of the tiger in Zinacantan, Chiapas, and the dance of the bulls in Huehuetenango, Guatemala.

Closely associated with the Deer Dance in Guatemala is the dog which plays an important part as assistant to the deer slayers. In Guatemala the masked dog dancer is called *La Maravilla* (the marvel) (Fig. 83). Among the Huichol Indians, who have retained many beliefs of their aboriginal religion, despite Christian incursion, a black dog is viewed as a companion of the dead and a guide to the underworld.

FLYING POLE DANCE, 84
DANZA DE LOS VOLADORES
Photograph by Grace Johnson

In this dance, which has been performed continuously through the centuries and is undoubtedly Precolumbian, the participants represent sacred birds who "fly" from a 100 foot pole to which they are attached by ropes, which may indicate the link between heaven and earth or the falling of the rain.

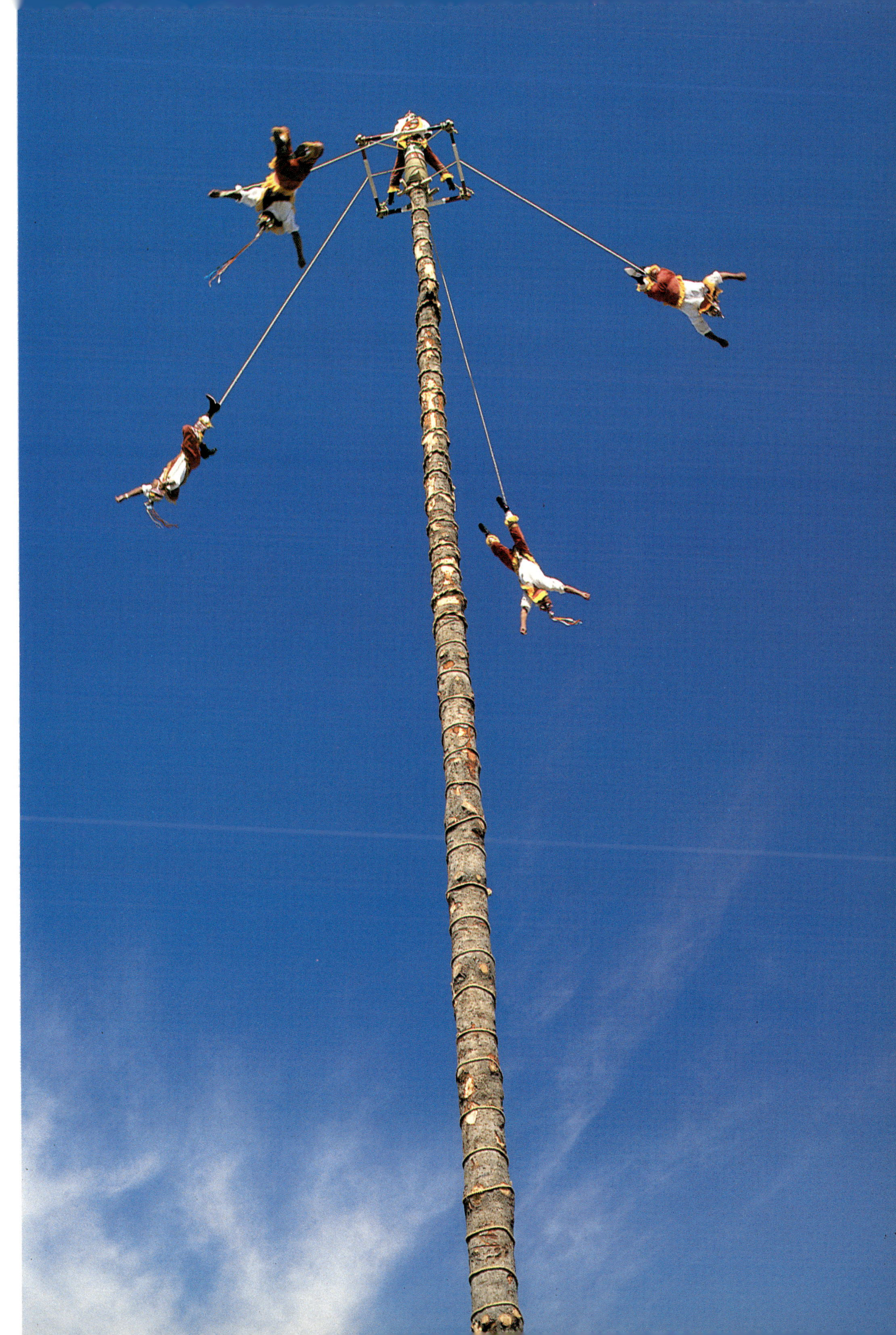

The deer is strongly linked to peyote (a cactus with hallucinatory properties) and to corn in the beliefs of the Huichol of Western Mexico (Fig. 82). In fact, the deer is the animal manifestation of peyote and considered most sacred. Traditionally, no important agricultural action could take place without a ceremonial deer hunt (Furst 1978: 23, 160).

The Flying Pole Dance (Los Voladores), Celestial Birds and Trees

Trees and poles (which are trees devoid of branches) figure in several of the "survival" dances. It is also clearly described in contemporary Maya mythology, as a giant ceiba (Bombax pentandra) growing in the center of the universe (Villa rojas 1967:275).

In the Flying Pole Dance. the participants sing, sway and balance on a tiny scaffolding 80–100 feet up, and then swoop like birds on ropes down to earth. It is probably the most well documented Precolumbian dance to have survived relatively unchanged from the time of the Conquest. Gibson (1964:151,504) reports that the Indians were ordered to perform it in 1571 at the 50th anniversary of the Conquest. Several Indian groups, including the Huasteca, Otomí, Nahua and Totonac speakers still perform this daring ritual (Fig. 84). Aboriginal elements in the dance include the ceremonial tree cutting, sexual abstinence for dancers, four dancers (five among the Otomí) representing birds who climb the pole and, after prayer and song by the captain, fly down the pole attached to ropes. They honor the old view of the universe by saluting the four directions. Frances Toor (1947: 319–320) in the early decades of this century, was told by an Otomí Indian that the dancers represented the four sacred birds that fly with the four winds to the cardinal points.

Until recently the dancers made exactly 13 revolutions around the pole to signify the 13 months of the native year. The captain, who remains aloft until the end of the dance, plays a small drum and flute in the native style. In earlier times, judging by the depictions of the dance at the time of the conquest and in the sixteenth century, dancers dressed in bird costumes (Illus. 11). Today most of the dancers are in Spanish influenced clothing with some bird feathers on their hats. In Pahuatlán, Puebla, dancers wear a feathered costume, but the revival is considered to have been generated by visiting anthropologists.

The bird-men in the dance could be linked to the four birds stationed at the quadrants of the universe They have been identified in the Codex Fejévary Mayer as the quetzal (East), yellow macaw (North), blue hummingbird (West) and a white sea bird (South) (Illus. 2 of previous essay). In Santa Cruz del Quiché, Guatemala, in the Dance of the Conquest, the Tecum Uman (Quiché King) wears a stuffed quetzal in his headdress (Frost pers.

comm. 1990). As this is a dance of Postconquest origin, the quetzal may signify native authority as opposed to the Spanish opponents. A bird with ritual and earth connections is the turkey, ancient symbol of Tezcatlipoca, the sorcerer god, which became a substitute for human sacrificial victims in native rituals after the Catholic Church forbade human sacrifice. In many areas of Mesoamerica, turkeys are sacrificed when a new home is built. A celestial bird connected with fertility in the ancient and contemporary Mesoamerica, is the hummingbird. It symbolized the young and virile warrior, and is still connected with love magic today.

Earth Creatures and the Snake Dances

The serpent in Mesoamerica has Prehispanic connotations today including those of rain and fertility. At the Preconquest feast of Atamalqualiztli (Eating of the Water Tamales) in honor of the rain god Tlaloc, the Aztecs performed a ritual at which Mazatec Indian dancers handled snakes and frogs and swallowed them alive as part of the performance. People fasted during this festival by eating nothing before noon for seven days, and after that only unsalted, unspiced tamales soaked in water. This was done once every eight years to give the corn a rest and revive it again (Sahagún 1950–1981, Bk. II: 177-178).

In Mexico, dancers handle live snakes during certain rituals. These dances usually include a female figure called a Malinche after Conquistador Hernán Cortés' mistress (in some areas she is called Maranguilla). In most cases the Malinche or Maranguilla is portrayed by a young man in women's clothing. In the Acatlaxquis dance in Pahuatlán, Hidalgo, which is considered to be Precolumbian in origin, the woman carries a gourd which contains a snake. At the end of the dance it is killed, symbolizing the defeat of evil. Gordon Frost (1976: 9–12) describes a snake dance (Danza de la Culebra) he observed and photographed in El Quiché, Guatemala (Fig. 85). The theme of fertility is predominant in his description of the sexual behavior of the dancers.

Serpents are connected with springs and waterholes and torrential rains among traditional Indian peoples of Mexico. The Yaqui and Mayo of Northwest Mexico and Mixe of Oaxaca share the ancient belief that this water serpent has horns like a deer. The Huichol of Jalisco and Nayarit believe in a horned serpent whose staffs gave support to Grandmother Growth (Berrin 1978: 144–145) (Fig 86 and illus. j). Serpents, as well as lizards and toads are often depicted in connection with a Europeanized devil (Fig. 87). This appears to be a combination of elements — the Christian idea of the devil as serpent in the Garden of Eden and earth linked animals from Preconquest times. The lizard (chintete) figures in the Dance of the Tenochtli (Dance of the Conquest) in Zitlala, Guerrero today, and has been identified with lust (Cordry 1981:34). The Malinche mask for this dance displays lizards biting her face (Fig. 88). There is still a strong negative feeling about this Indian woman who helped Cortés in the Conquest of Mexico, through her knowledge of Indian languages, her loyalty to him, and her diplomatic skills.

Yet another serpent, the mythical plumed serpent of Preconquest belief has withstood time in Mesoamerica. He is honored at the Feast of San Sebastián by the Tzotzil Maya. The headdress worn by the plumed serpent dancers at the festival have features of the Quetzalcoatl of old, including a conical spotted hat and the beak mask. Hunt (1977: 76–77) interprets his appearance to represent the fertility of the land and the crops in the post-harvest period.

The armadillo, or "turtle-rabbit" of the Aztecs, is widespread throughout Mesoamerica and connected with the earth and fertility (Fig. 92). Lumholtz, in his travels and studies among the Huichol Indians at the turn of the century, reported that they believed that the armadillo was the husband of the earth goddess, the oldest woman in the world (Lumholtz 1902:165). The animal's connection with fertility and the earth is possibly reflected in the use of armadillo seed bags for the sowing of corn in Chenalho and Huistán, Chiapas and Yalalag, Oaxaca.

Stuffed furry animals such as squirrels, racoons and coatis are used by clowns in Guerrero and a racoon tail wand is carried by Huichol Indian clowns. A stuffed squirrel with red-painted genitals is used as a sexual symbol during the buffoonery that takes place during the feast of Saint Sebastián at Zinacantan, Chiapas (Vogt 1969:542).

Dance of the Jaguar or Tigre

Most powerful and fearsome of the Mesoamerican mammals, the jaguar is featured in dances throughout the southern area (Fig. 89) as well as Guerrero and Morelos, Mexico. The dance-ritual is basically Precolumbian (Markman and Markman 1989:167) (Fig. 90). Common features are a simulated hunt, ritual combat and use of a full body costume and large jaguar mask. Known as a tiger *(tigre),* the jaguar also appears in carnival dances in the Costa Chica of Oaxaca, where the tiger becomes a lewd performer in a real life drama that may end in violence (fights) and even death (Flanet, in Markman and Markman 1989:173). The costume and combat aspects are reminiscent of the Aztec jaguar knight's array and behavior at the Aztec feast of Tlacaxipehualiztli (the Feast of the Flaying of Men). In this Precolumbian ceremony the jaguar skin clad warrior armed with a sharp obsidian bladed weapon waged a one sided battle with a captive tied to a heavy stone wielding only a wooden club. The battle ended with the defeat and sacrifice of the captive.

Today the dance in Zitlala, Guerrero, includes a battle with heavy knotted leather whips and ends with the combatants gravely wounded. The combat was undertaken so that the rains would come, linking the jaguar to fertility. The Markmans (1989: 171–178) believe that the jaguar mask has a history dating back to Olmec times three thousand years ago and that "the symbolic meanings associated with that metaphoric disguise have remained constant." That is, these jaguar dancers, by their performance, and personal sacrifice are still assuring that the rains will come in sufficient quantity and in time to produce a bountiful harvest. Jaguars are also associated with other natural phenomena. The Tzotzil believe that earthquakes are caused by a giant jaguar scratching himself, and some Chol Indians attribute eclipses to the attack on the moon by a cosmic jaguar. These beliefs relate to the Preconquest view of the jaguar as related to disaster, the power of evil and the god Tezcatlipoca (Smoking Mirror).

Flowers as Motif and Medium

Flowers were widely revered and used in Precontact times as decorations for gods, people and objects. Certain flowers retain ancient meanings. The marigold *(tagetes)* is the predominate ceremonial flower today and plays an indispensable part in the celebration of the departed souls on the Days of the Dead. A small flower called *toto* by the Huichol Indians is important in their art, especially weaving and embroidery but is also used as a design motif in beaded gourds. It is a symbol for corn and peyote (Berrin 1978:184). Among the Yaqui and Mayo, flowers are a metaphor for the life-force in the enchanted forest *(huya ania)* of the native religion and an indispensable part of the rituals associated with it (Markman and Markman 1989:189).

Today in rural villages, artists create processional floats made of seeds and flowers to honor patron saints. Flower petal carpets are assembled, and the saint images are draped with garlands (and clothing) much as were the ancient gods. Colorful and fragrant floral decorations adorn home altars on the feast of the Days of the Dead and the Virgin of Sorrows. Dried flowers are used also for ceremonial paraphernalia and for Christmas decor in Oaxaca. In San Antonino Castillo Velasco, Oaxaca, a Zapotec village known for its intricate embroideries, the artisans also fashion dolls made of dried flowers during the Christmas season (Fig. 91).

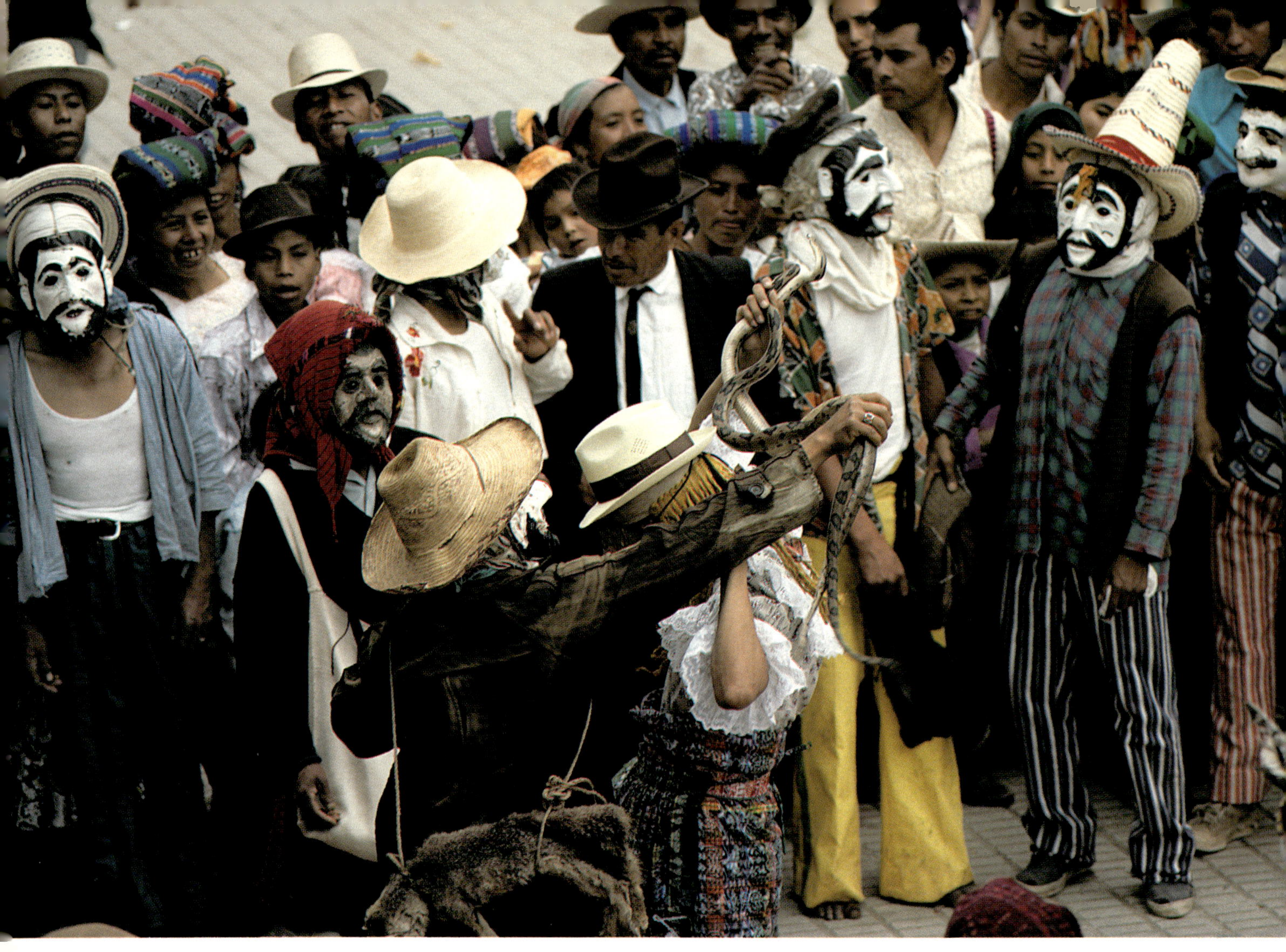

SNAKE DANCER IN THE   85
DANCE OF THE SERPENTS

Quiche Maya Indian, Santa Cruz del Quiche,
Guatemala
Collected in 1975
Gordon Frost Folk Art Collection, Benicia.

In this dance *(La Danza de la Culebra),* men dressed as
women handle live serpents during the dance which is
related to fertility and penance.

SERPENT STAFF OF   86
GRANDMOTHER GROWTH

Huichol Indian, Jalisco/Nayarit, Mexico
Wood. Measurements not available
Identified by Peter Furst
Collected in 1968
The Natural History Museum of Los Angeles County
A-8400-70-56.

GRANDMOTHER GROWTH   j
WITH SERPENT STAFF

Huichol Indian deity, West Mexico. (Adapted from
Lumholtz 1902, II: 163)

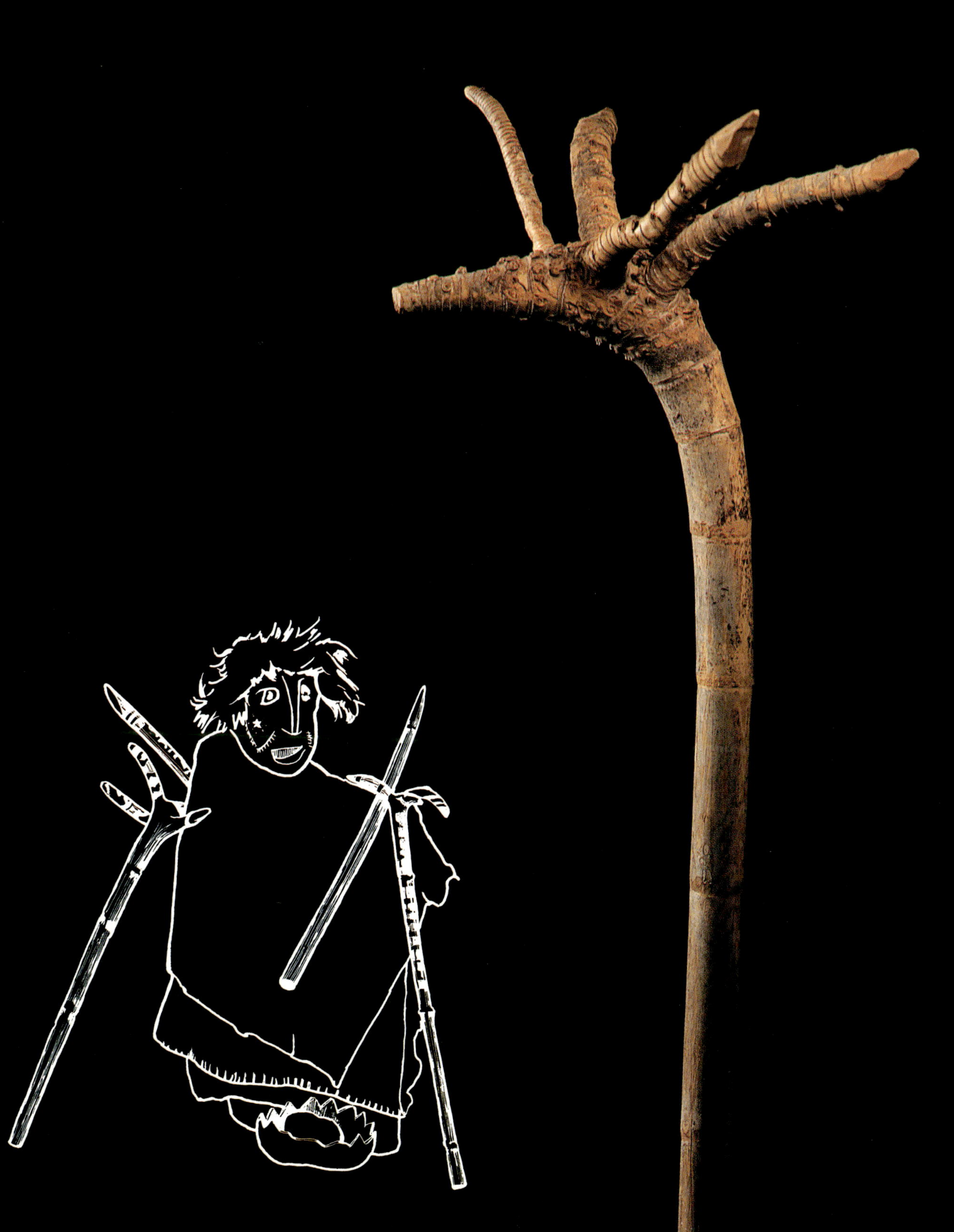

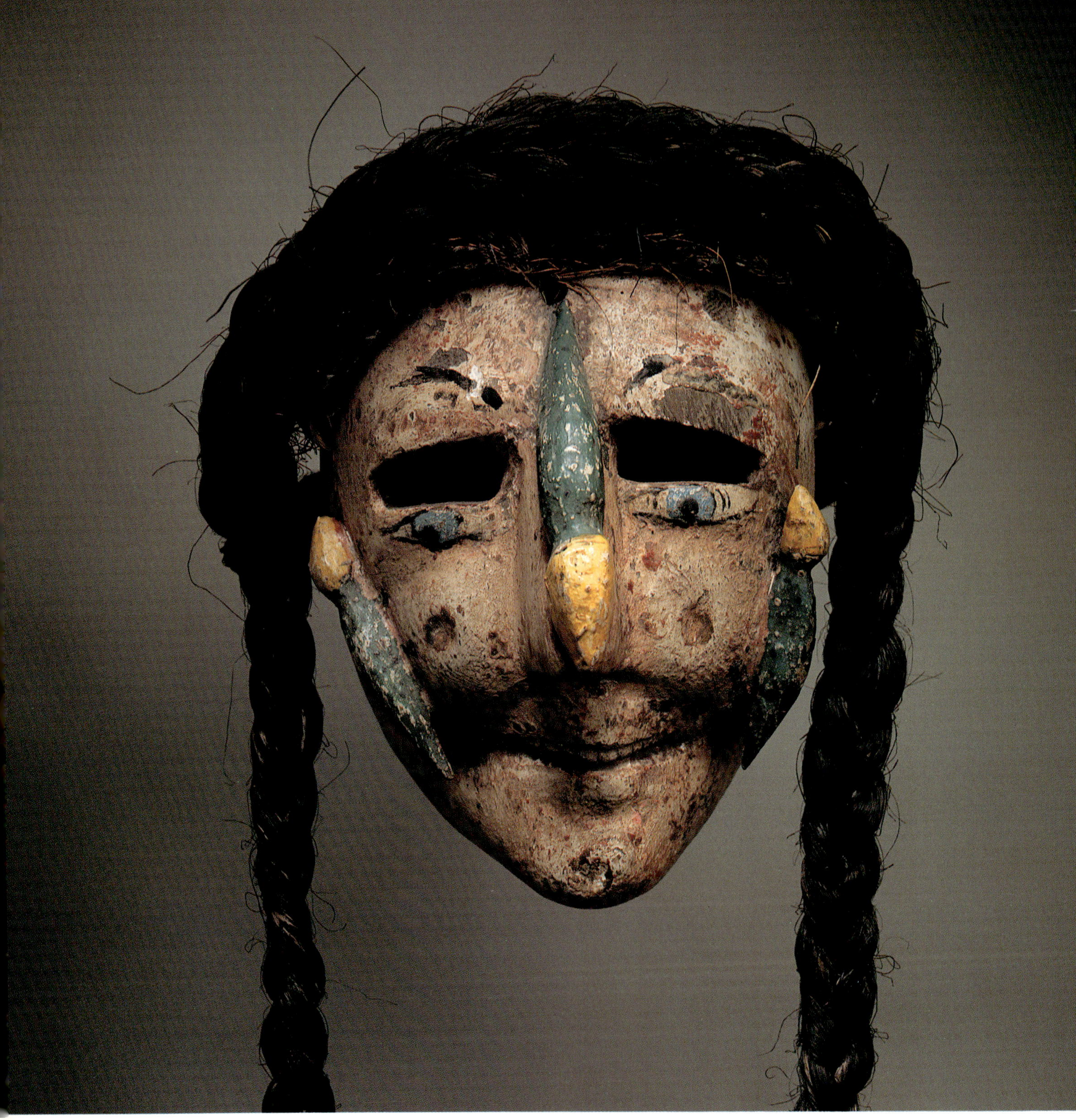

MALINCHE MASK,   88
PROBABLY DANCE OF THE TENOCHTLI
Nahua, Zitlala, Guerrero
13″
Collected circa 1980
Tom and Alma Pirazzini Collection

Stinging lizards known as chintete are found on the cheeks of the mask which has been interpreted to signify lust or wantonness on the part of the woman who was Conquistador Hernan Cortes' mistress and interpreter.

BLACK DEVIL MASK   87
Cakchiquel Maya, Sacatepequez, Guatemala
Wood. 10″ × 16″
Collected in 1974
Gordon Frost Folk Art Collection, Benicia. 7F412412

This mask (known as the *Diablo Mayor,* or chief devil) with serpents, toads and lizards on the face was used in the Dance of the Devils *(Baile de los Diablos).* The traditional underworld creatures of the ancient world are blended into the mask of the Christian Satan.

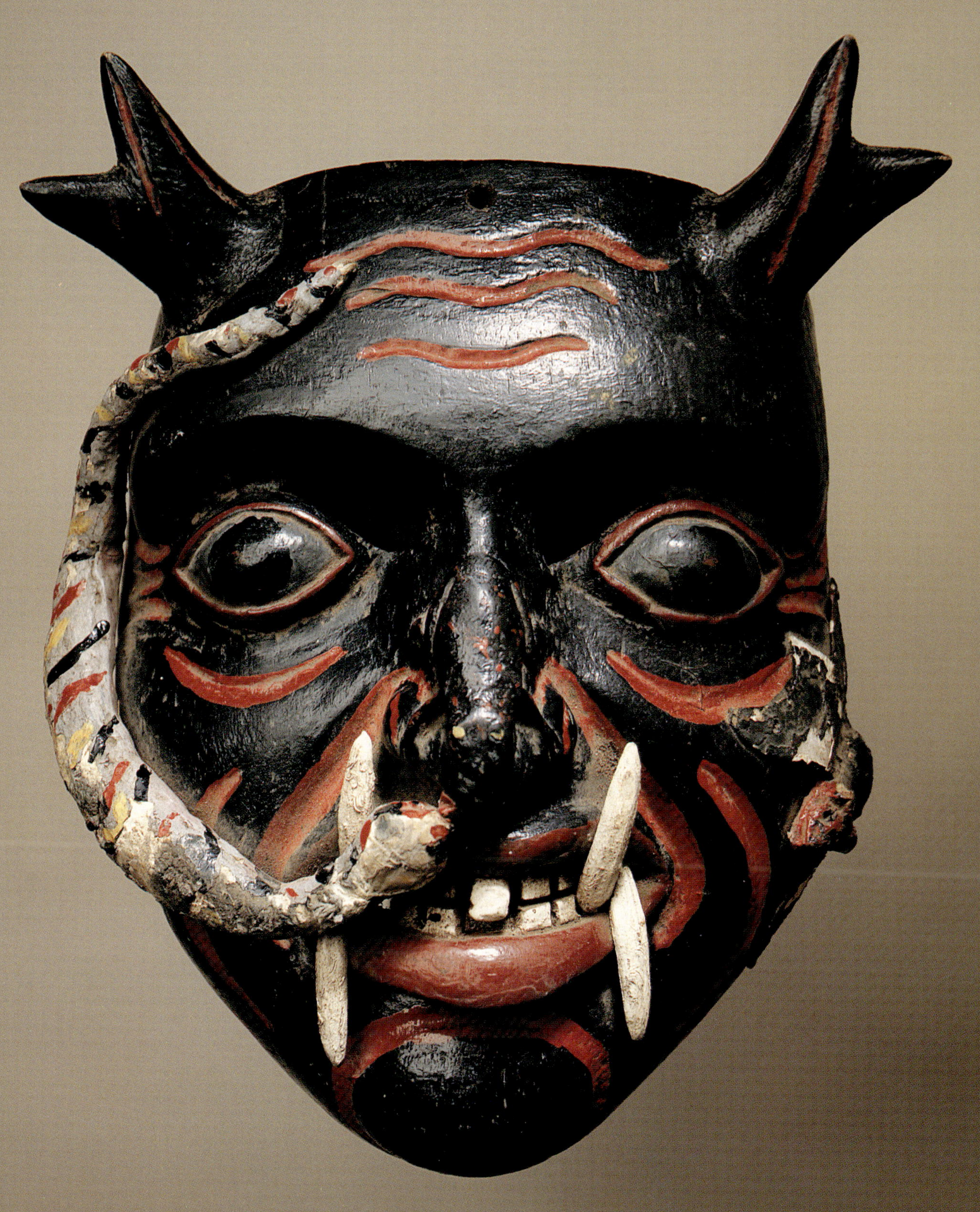

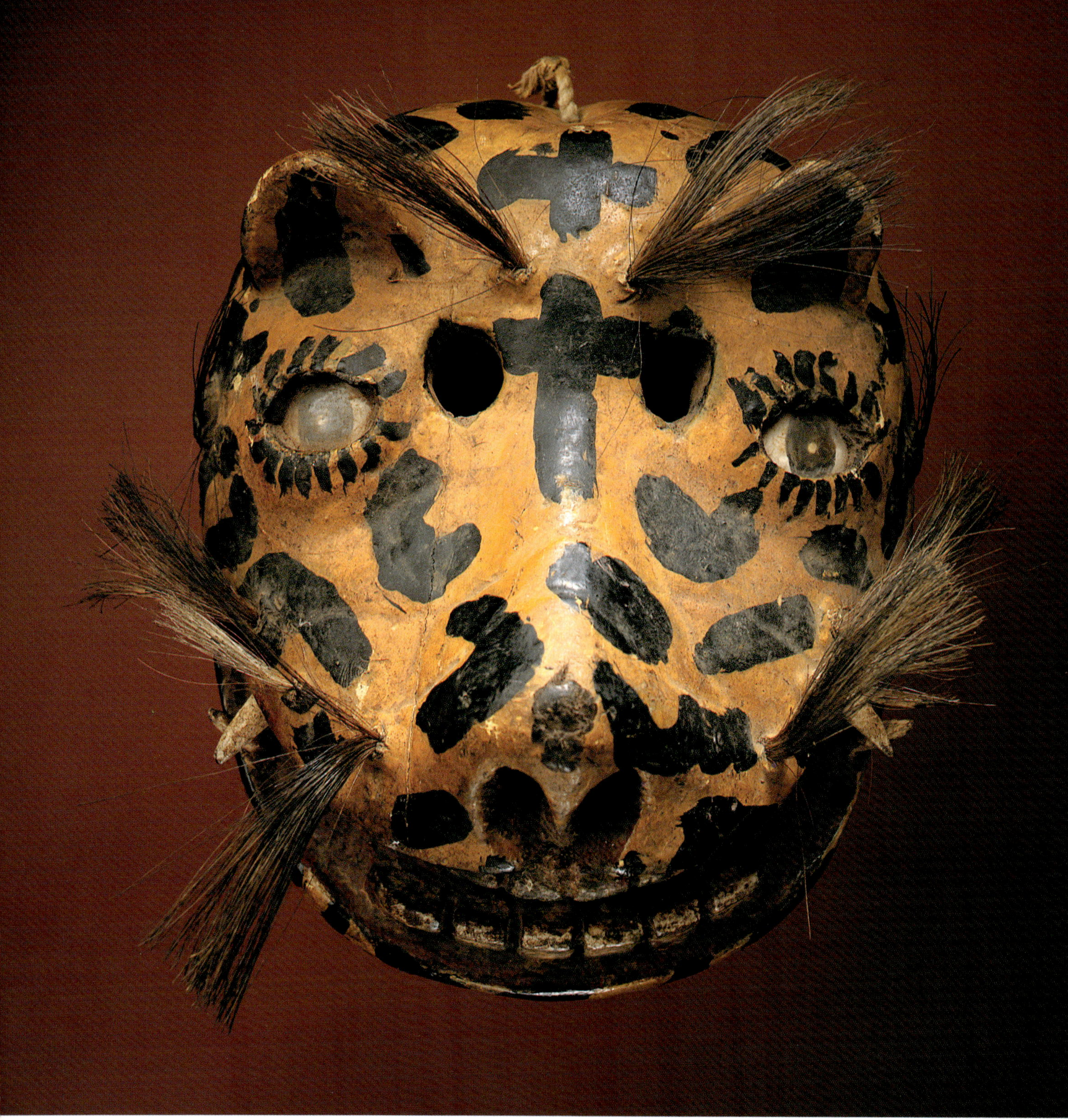

SPOTTED JAGUAR MASK 89
Maya Indian, San Marcos, Guatemala
Wood. 7½″ × 9″
Collected in 1974
Gordon Frost Folk Art Collection, Benicia. 7F355

This mask was used in the Deer Dance, *El Baile del Venado*. The jaguar symbolizes rain and fertility in animal dances in several regions of Mesoamerica. This mask has two crosses on the face that indicate Christian influence.

JAGUAR DANCER PERFORMING IN THE DEER DANCE 90
Mam Maya Indian, Todos Santos Cuchumatanes, Huehuetenango, Guatemala
Photographed in 1978
Gordon Frost Folk Art Collection, Benicia.

The jaguar appears as a regular character in the deer dance in Guatemala. His association with fertility is well documented.

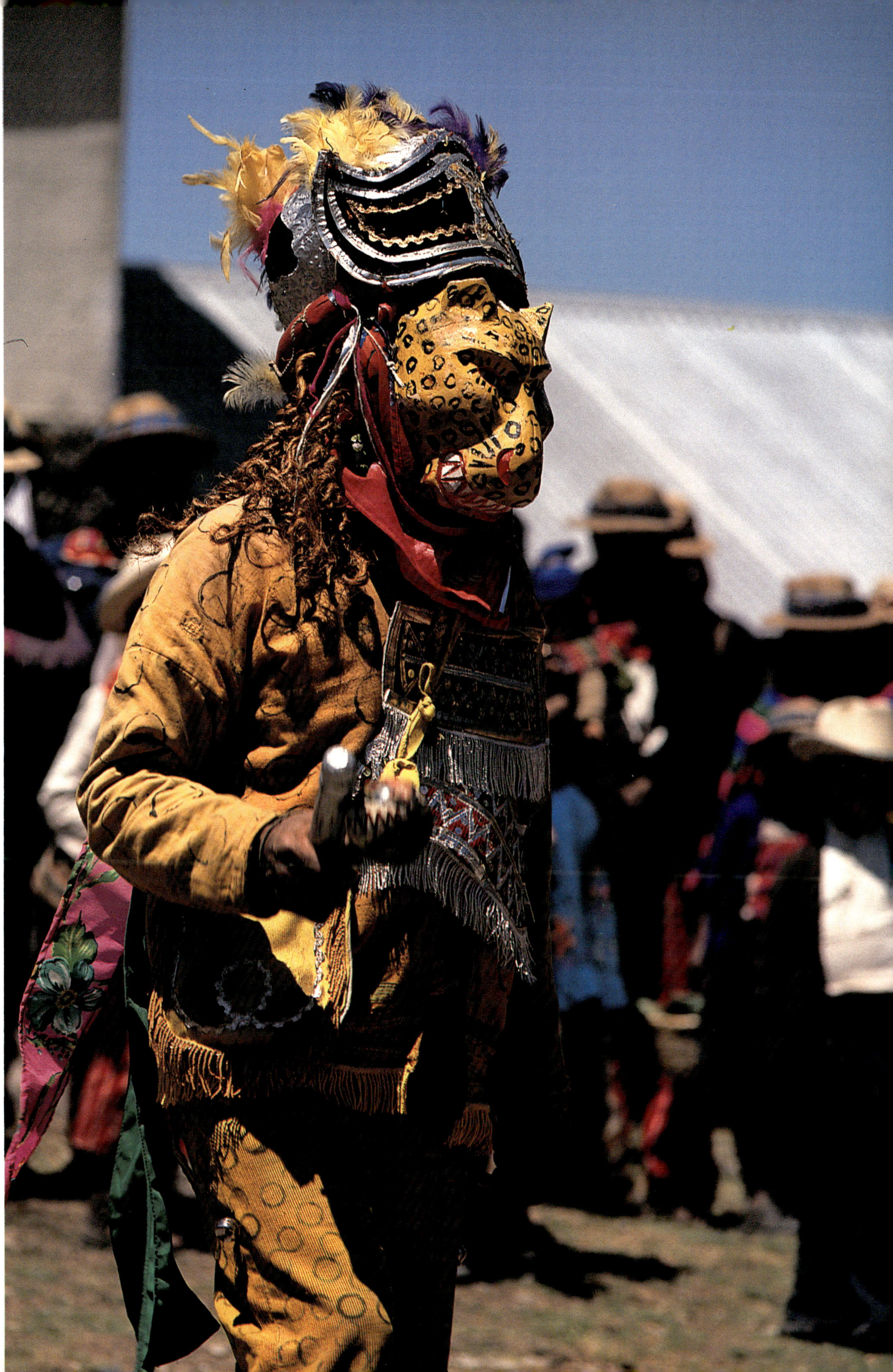

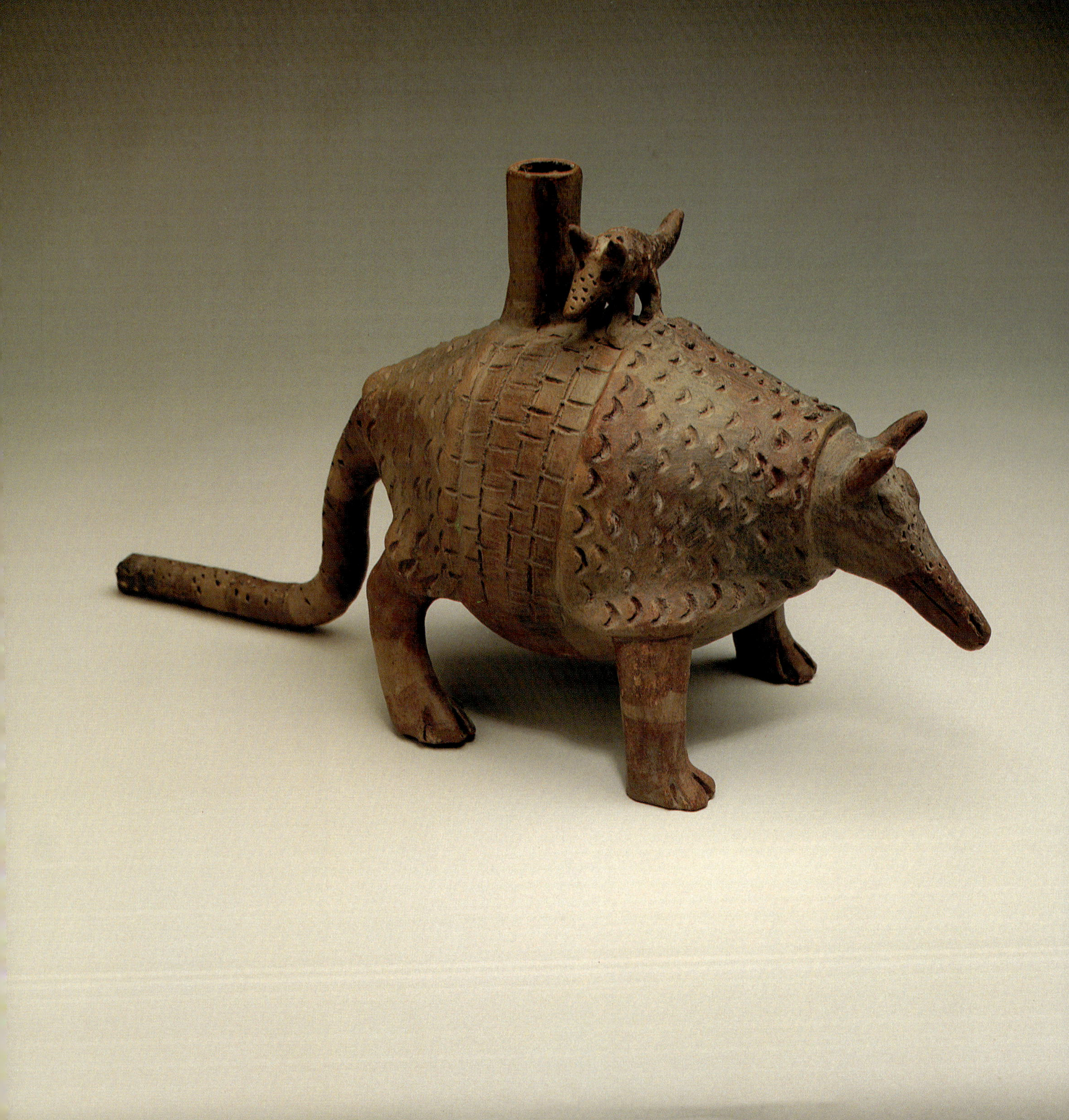

ARMADILLO CANDLEHOLDER   92
WITH BABY ON ITS BACK
Mexican, Acatlán, Puebla, Mexico
Fired clay, unpainted. L: 16″
Katarina Real Collection A-125

In Precolumbian times the armadillo was associated with
earth and fertility.

TWO FEMALE FIGURINES   91
OF DRIED FLOWERS
Probably Zapotec Indian, made in San Antonino
Castillo Velasco, Oaxaca, Mexico
Dried flowers on a base of avocado leaves. H: 12½″
Made in December 1986
Tom and Alma Pirazzini collection

These figures are made and sold at Christmastime in the
region and wear traditional headdresses.

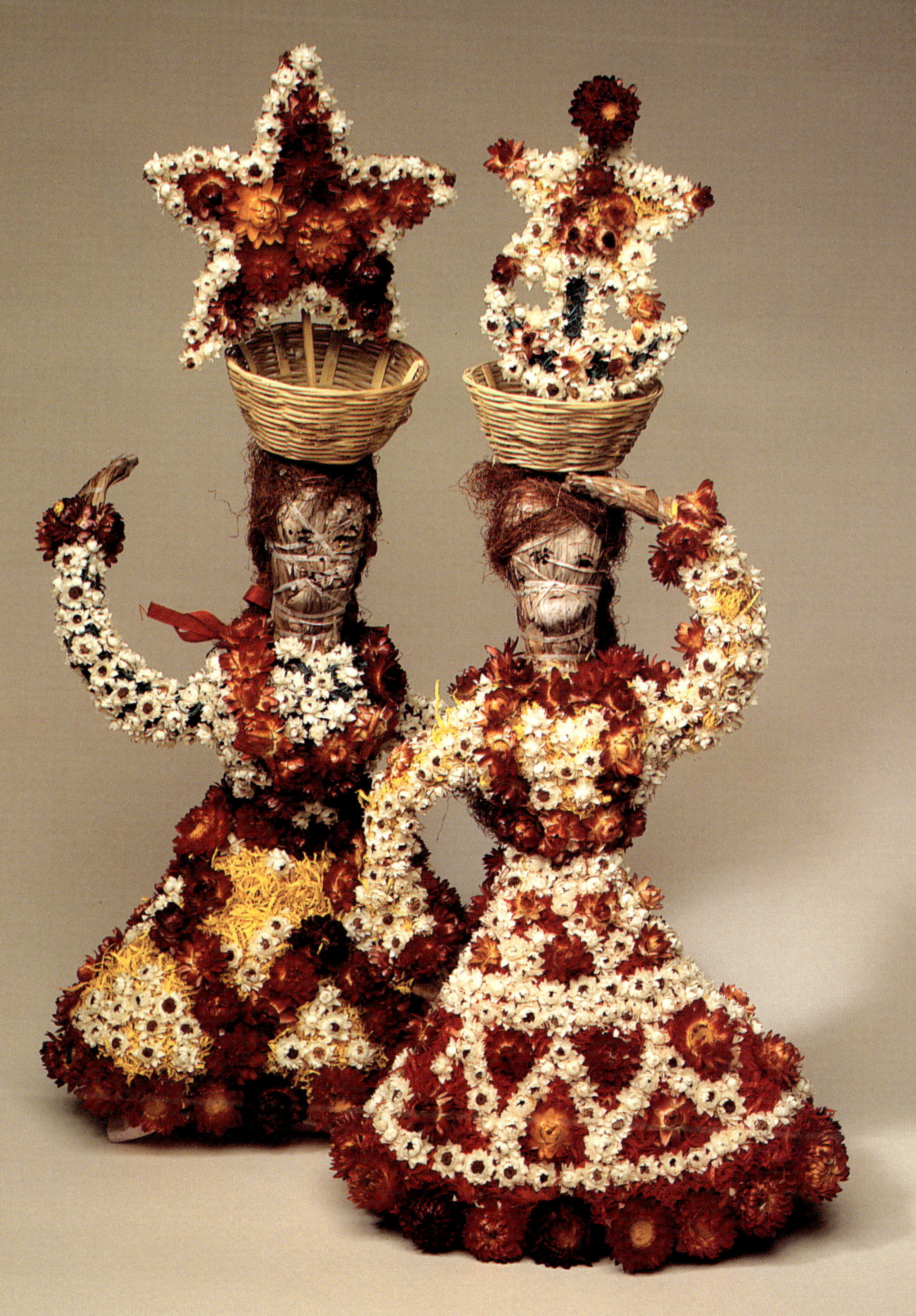

SYNCRETIC THEMES USING FLORA AND FAUNA

Many rituals and the artifacts connected with them are a blend of the native Indian and European elements. Whereas even those previously listed as native survivals have some post-conquest themes (i.e. the *volador* costumes), others have a greater proportion of Christian elements.

The Cosmic Tree, the Cross, and the Tree of Life

The basic form of the Precolumbian universe included a cosmic tree in the "navel of the earth" penetrating the layers of upperworld and underworld. The Lacandon Maya of Chiapas believe the sun passes through to the underworld each evening by descending through the roots of a tree. The contemporary Tzotzil Maya view the sky as a mountain with 13 steps surrounding a gigantic central ceiba tree (León-Portilla 1973:140).

It served the Catholic Church well that the world tree resembled a cross. However, in many rural areas today, it is the cross itself rather than the figure of the crucified Christ that is the principle object of worship. House crosses which adorn the roofs or patio entrances are known from Southern Mexico particularly, including Oaxaca (Fig. 78) and Chiapas. In Zinacantan, Chiapas the crosses symbolize the people living in one household and are the doorway to the "soul" of the house (Vogt 1969: 127–128).

Trees are important in several animal dances. In the dance of the bulls in Huehuetenango, Guatemala, trees are hung with fruit. A similar ritual is noted in Prehispanic times for the Aztec new year during the feast which honored the rain god Tlaloc when trees were adorned with corn and fruit to give thanks for abundant harvests. A cross adorned with ears of corn was found in 1970 in the Western Highlands of Guatemala.

Trees in the Prehispanic world correlated with the abundance of food and progeny. The colorful earthenware trees of life made today in Mexican villages may relate to an ancestral tree. Adam and Eve are often the main figures at the base of the tree of life and it was traditionally given as a wedding gift in Puebla, which gives some credence to the idea. The oldest of the existing clay trees of life appear to be from Izucar de Matamoros, Puebla. To my knowledge, there are no historical descriptions or extant colonial examples. So, while an intriguing idea, the Precolumbian connection is relatively uninvestigated. The making of trees of life in ceramics apparently moved from Huaquechula, Puebla to Izucar de Matamoros (Fig. 93) and later to Acatlán, Puebla and Metepec, in the State of Mexico, where the majority are made today. Many traditional ones have the Adam and Eve theme, but artists more recently combine Christmas creches *(nacimientos)* and other motifs such as the mermaid or a variety of animals.

The Horse of St. James (Santiago)

The festivals dedicated to St. James are examples of devotions by many indigenous groups to an imported patron saint. The saint of the village, dressed, bejeweled and garlanded with flowers, is the object of reverence and local possessiveness, much as the village deity was in Precolumbian times. The dance rituals once devoted to pre-Christian deities are now performed in honor of the village saint. St. James, or Santiago, is the patron of Spain, a warrior saint who fought the Moors. He is depicted as a bearded warrior astride a horse and wielding a sword.

During the dance in Cuetzalan, Puebla the dancer portraying Santiago wears a jointed wooden horse attached to his body so he appears to be riding. The most interesting aspect is the veneration bestowed upon the wooden horse itself (Fig. 95). At Cuetzalan, the horse is passed on to the next dancer representing St. James to care for until the following festival. He feeds it corn and water daily.

In San Juan Peyután, Sierra de Nayarit, Mexico, a St. James is defeated by the pagan enemy, which is clearly a native innovation to the Spanish dance (Toor 1947:350). This is one of many "reversals" in ritual which the Indians created in adapting Spanish elements into their ritual system.

The Days of the Dead

The Days of the Dead *(Dias de todos Muertos)* are a widespread syncretic observance throughout Mexico. While nominally a Catholic celebration of the traditional Day of All Saints, Nov. 1, and Day of All Souls, Nov. 2, it contains pre-Christian elements. The Aztec flower of the dead, the marigold *(tagetes)* is used on homemade altars constructed to honor the family's deceased members. Food and drink, as well as favorite articles (i.e., children's toys) are placed there for the visit of the spirits. People also carefully fashion altars at gravesites. There is a tradition of mocking humour mixed with wholehearted acceptance of the returning souls of the dead on these days.

TREE OF LIFE, *ARBOL DE LA VIDA*   93
Mexican, Izucar de Matamoros, Puebla, Mexico
Fired earthenware. H: 38″
Collected in 1967–8
San Diego Museum of Man 1973-22-1

This is the traditional type of Tree of Life from Puebla which was given as a wedding gift and may be related to the ancestral trees illustrated in the painted books or codices of ancient Mesoamerica, although Christian angels are also part of its ornamentation.

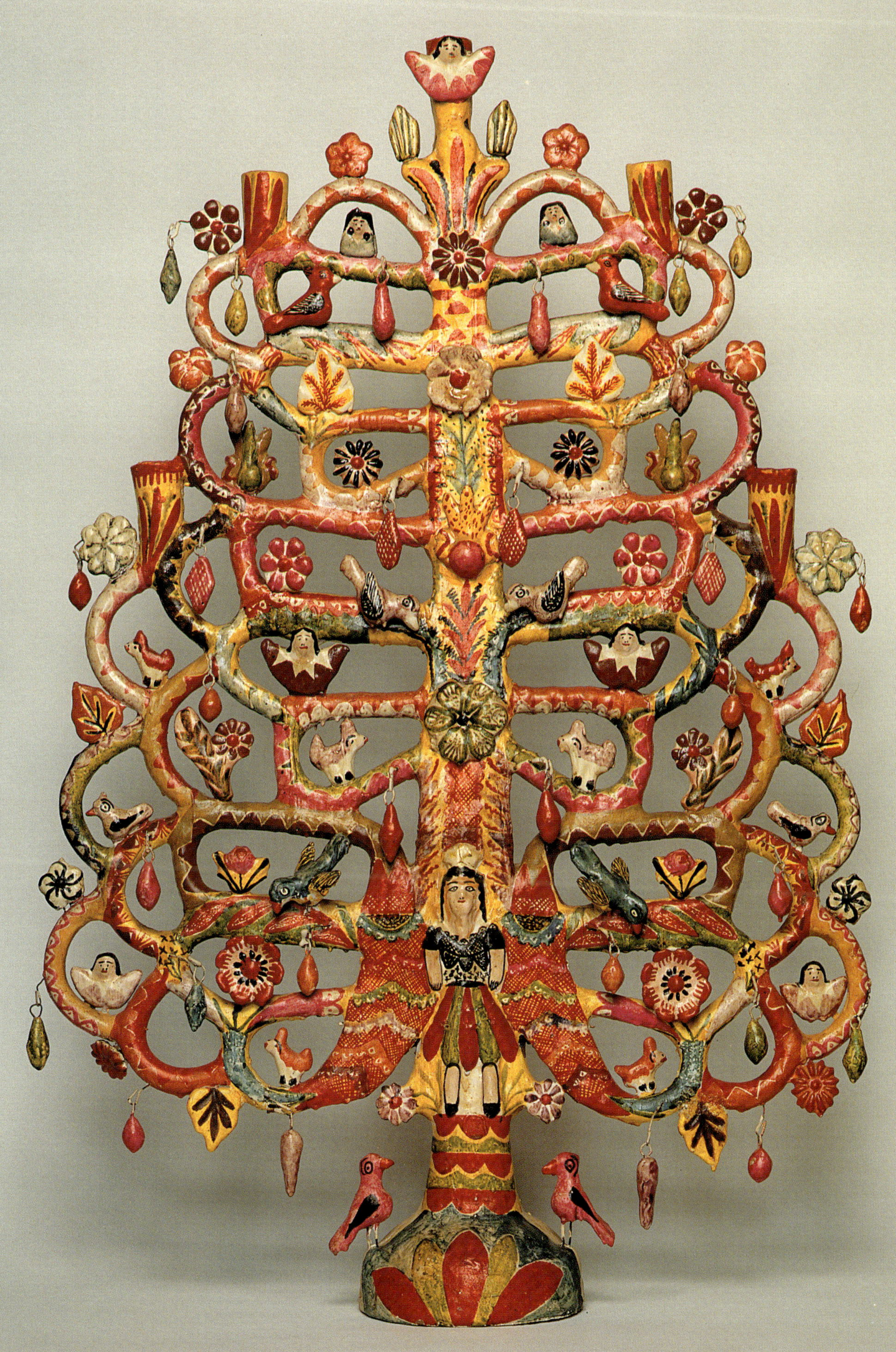

The Aztecs had several customs paralleling those practiced today. Food, flowers and incense were annually offered to the dead until the fourth anniversary of the death which signalled the end of their journey. Those who attained one of the heavens, due to their manner of death,were honored at festivals. One of these celebrations was Tepeihuitl (Feast of the Hill) in the thirteenth month of their calendar, at which people consumed images made of amaranth dough. This custom possibly assisted in the acceptance of the Spanish one of baking and distributing "Bread of the Dead" on the Days of the Dead (Green 1972:259).

Post-Conquest Domesticated Animals

The Indians had few types of domesticated animals before the Conquest. The dog, the turkey, the bee and possibly the dove were known domesticates. The European imports included cattle, horses, donkeys, pigs, chickens, sheep and goats as well as the household cat.

These newly introduced animals were incorporated into religious and secular ceremonies and portrayed in ritual art. In some cases they substituted for the human beings or animals formerly sacrificed. Hens were used to replace the scarcer turkeys in animal sacrificial rites. Chickens and other fowl decorate the ceramic coffee vessels which are important parts of the *cofradía* (religious organizations) rituals in Guatemala. The horse is honored in the Santiago dances and widely used as an art theme. Sheep (Fig. 94) and donkeys became part of the complex of *nacimiento* (Christmas creche) animals.. During the Days of the Dead, in Toluca, State of Mexico, artisans make the traditional sugar sheep for altar offerings. Goats, too, have a ceremonial role among the Yaqui of Northwest Mexico as the one type of animal mask used by the Pascola dancers. The Markmans (1989:188) have presented a case for the Pascola ritual maskers and dancers as a syncretic one in Mesoamerica.

Cattle were a welcome Spanish introduction to the New World, since there were no beasts of burden in Mesoamerica prior to the Conquest. The bull took on a ceremonial role. He is to be found in the bull dances of Guatemala (Fig. 96), bull sacrifices among the Huichol, ritualized bullfights in Chiapas and ritual objects such as the incense burners used by the Mayo and Yaqui of Northwestern Mexico. The bull figures in a number of dances today as a powerful figure and a match for the strongest of men. Bull dancers appear in carnival celebrations in Jamiltepec, Oaxaca, Mexico in context with dancers called *rubios* (blonds) in cowboy attire (Brown 1989:254). This context is similar to that of the bull dancer in the Dance of the Mexicans (El Baile de los Mejicanos) in Sololá, Guatemala, where he appears with characters dressed as cowboys. The bull's appearance sometimes signals Spanish associations such as the bullfights and fireworks. The *torito* is a wooden frame in

the shape of a bull's head laced with firecrackers and set atop a man's back. It is eventually ignited to the delight of fiesta-goers. The symbol of the fireworks specialist in rural Mexico is a bull's head effigy outside the workshop.

NATURALISTIC THEMES NOT KNOWN TO BE SURVIVALS

Mesoamerican artists have produced a wealth of works using the animals and plants of their surroundings, with no overt symbolic significance. Purely utilitarian objects such as bean pots, casseroles, mole dishes, chocolate pots and wooden chocolate beaters, horn combs, toys, and miniatures, are often decorated with, or made in the form of, flora and fauna. In some cases, naturalistic representations were encouraged by outsiders, or simply started by a particular artist.

This category comprises the great bulk of Mesoamerican folk art produced today. Many books have been written about the classification, type-styles, and regional differentiation in Mesoamerican folk art. Some of this art is created for village use or home use, like the woven and embroidered clothing, kitchen utensils, toys for children, wooden chests, and miniatures. A simple but strong dog image in the form of a bench was collected in San Pablito, Puebla, an Otomí Indian village, and typifies folk art in use by the people who created it (Fig. 97).

Artists of special talents, marketing skills, or outside recognition, have emerged in recent years, such as Teodora Blanco of Atzompa, (Fig. 98), and Doña Rosa Real de Nieto of Coyotepec, Oaxaca., José Limón and Amado Galván of Tonalá, Jalisco (Fig. 99), Heron Martinez of Acatlan, Puebla and Mónico Soteno of Metepec in the State of Mexico. With wider recognition through word of mouth, exhibits and publications , variety in many areas has expanded. The craft inventory of the traditional pottery village of Tonalá, Jalisco has expanded its repertoire of animals to a variety worthy of Noah's Ark. Some of the old-style burnished ware (Fig. 100) and *petatillo* ware retains the techniques and style developed during the colonial period. The pottery figurine makers of Tlaquepaque (especially the well known Panduro family) still fashion detailed everyday life scenes of painted or unpainted clay.

Research on survivals of Prehispanic beliefs in Mesoamerica has burgeoned in recent decades. Scholarly investigations encompass more than flora and fauna themes, but given the close association between man and nature in Mesoamerica, these motifs are involved. Investigations continue to mutually illuminate the Precolumbian world and our understanding of the contemporary Mesoamerican spritual beliefs as expressed in their art.

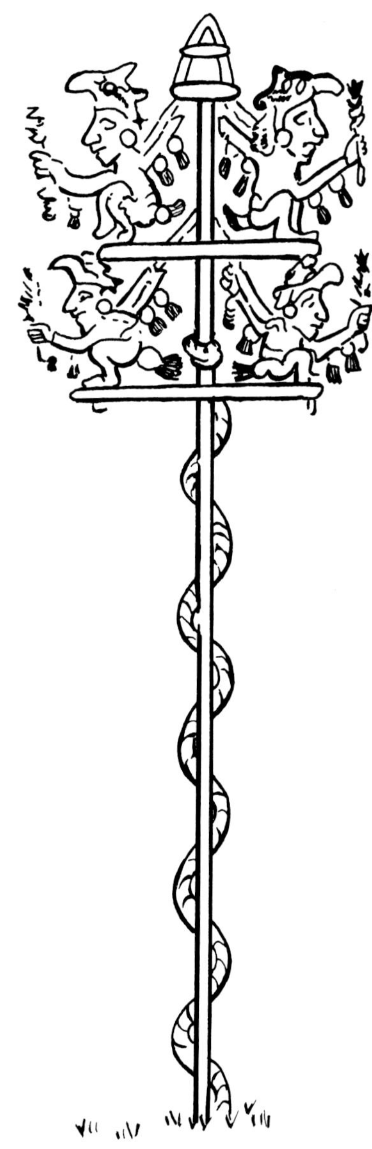

END NOTES

1. This essay is not a comprehensive review or even a summary of the survivals and syncretic elements in folk art which have been studied to date in Mesoamerica. Building on the foundation established by such early researchers as Carl Lumholtz (1901–1902), Frederick Starr (1899, 1900, 1902), Robert Redfield (1934), Frances Toor (1947) to the ethnographic works published in the Handbook of Middle American Indians edited by Robert Wauchope (especially vols 6, 7, 8) (1964–1976) and the works of Evon Vogt (1969, 1976), Peter Furst (1972), Eva Hunt (1977), Ted Leyenaar (1978), Gary Gossen (1986), Barbara (1982) and Dennis Tedlock (1985), Janet Esser (1988), Roberta and Peter Markman (1989) and many others, an array of survivals and syncretic elements, some with flora and fauna themes, have been identified. For those who would become more familiar with the subject the following bibliography is recommended.

2. For readers interested in textiles, as we had to leave this rich source of animal and plant themes out of our essays, a good source of information on the subject can be found in Irmgard Weitlaner Johnson's works and bibliography (1971, 1976).

3. A wedding huipil from Zinacantan, Chiapas, which has chicken feathers incorporated in the weaving, is in the permanent collection of Mexican costumes at the San Diego Museum of Man (Cat. 1987-54-1).

VOLADOR RITUAL   k
OR FLYING POLE CEREMONY
Four birdmen at top poised to fly 16th century *Codex Fernandez Leal* (Adapted from Esser 1988: Fig. 119)

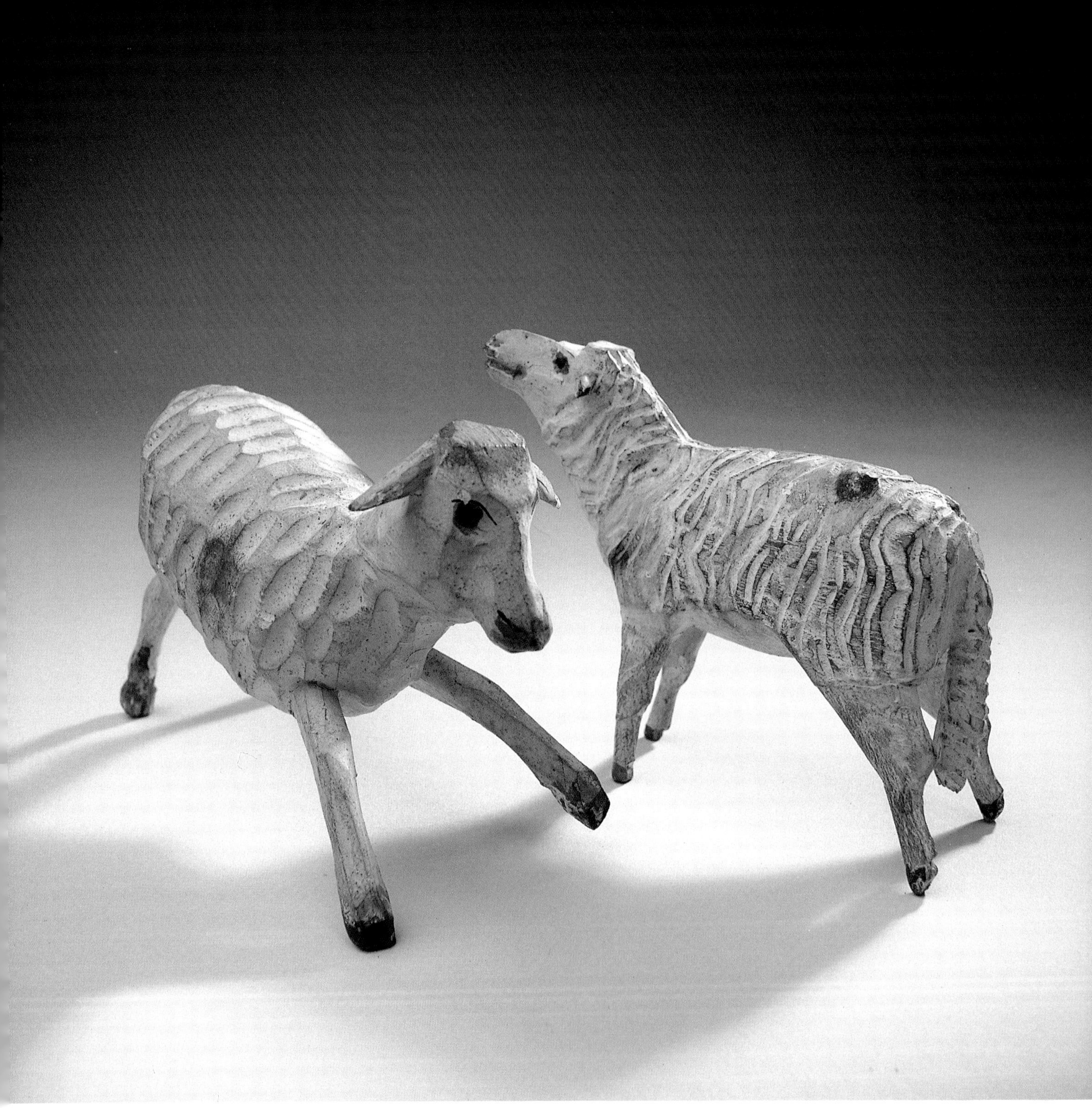

SHEEP   94
Mexican, provenance
Wooden. One figure H: 3½″. Second figure H: 4″
Collected in 1989
Mingei International Museum of World Folk Art

HORSE OF ST. JAMES (SANTIAGO)   95
Attributed to Nahua Indians, Landa (?), Guerrero, Mexico
Wood, cloth, horsehair and leather.   L: 19″
Collected in the early 1980's
Tom and Alma Pirazzini Collection

The horse is jointed and fitted with clasps to hold it to the
body of the dancer who is selected to be St. James each
year.

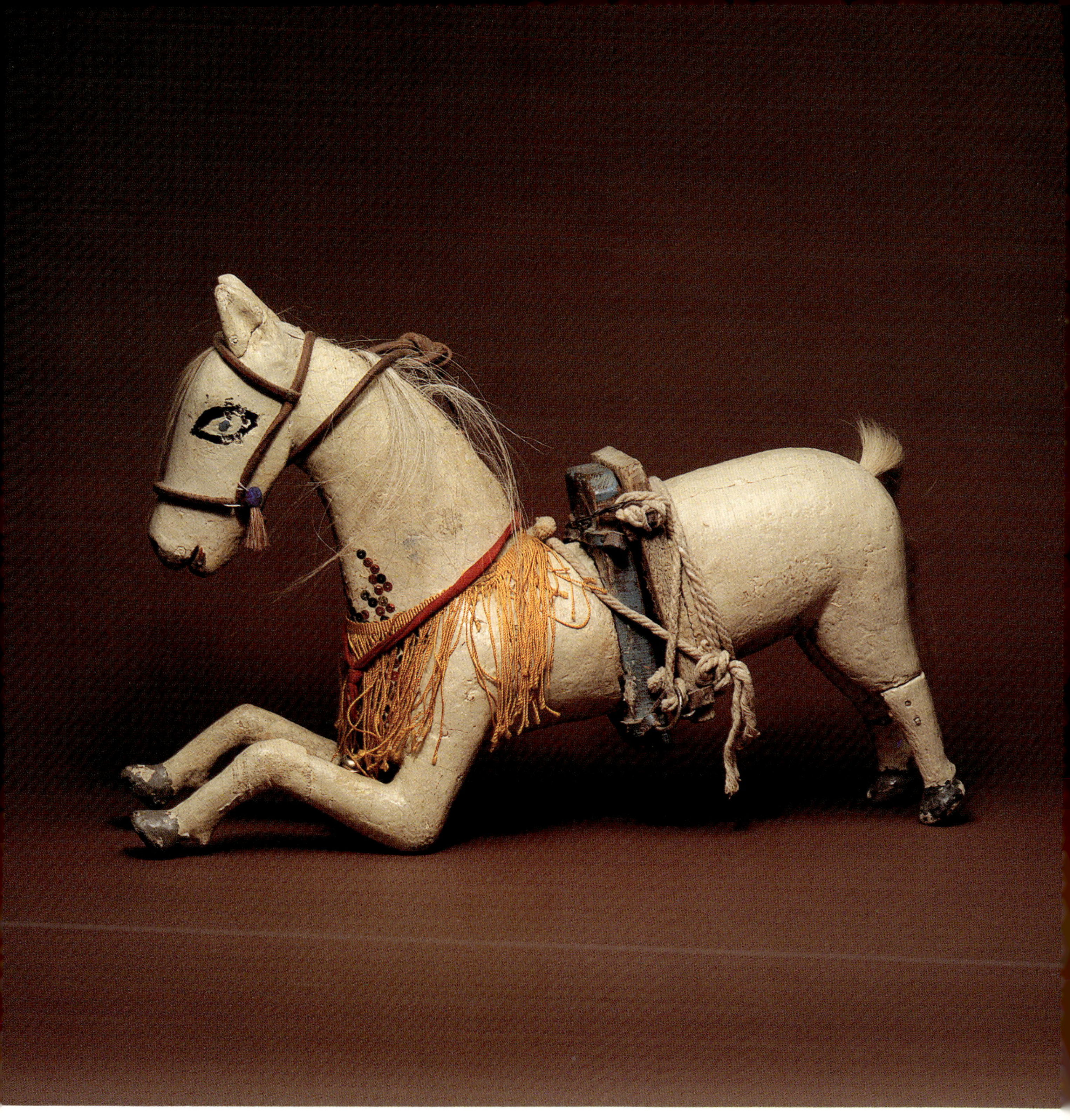

SELECTED BIBLIOGRAPHY

Adams, Richard N. and Arthur J. Rubel
1967   "Sickness and Social Relations," in *Handbook of Middle American Indians*, (Robert Wauchope, gen. ed.), vol. 6, pp. 333–355. University of Texas Press, Austin.

Aguilera, Carmen
1985   *Flora y Fauna Mexicana: mitología y tradiciones*. Ed. Everest Mexicana, S.A., Mexico.

Altman, Ralph
1961   "Comments," *Current Anthropology*, vol. 2, No. 4, pp. 355–358.

Bardawil, Lawrence W.
1976   "The Principal Bird Deity in Maya Art: An Iconographic Study of Form and Meaning," in *The Art, Iconography and Dynastic History of Palenque, Part III* (Merle Green Robertson, ed.), pp. 195–209. The Robert Louis Stevenson School, Pebble Beach, California.

Beals, Ralph M.
1945   *The Contemporary Culture of the Cahita Indians*. Bulletin 142. The Smithsonian Institution, Bureau of American Ethnology, Washington, D.C.

Benson, Elizabeth P.
1985   "The Classic Maya Uses of Jaguar Accessories," in *Fourth Palenque Round Table*, 1980 (M.G. Robertson, general ed. and E.P. Benson, vol. ed.), vol. VI, pp. 155–158. Pre-Columbian Art Research Institute, San Francisco.
1988a   "The Maya and the Bat," *Latin American Indian Literatures Journal*, 4:2, pp. 99–124.
1988b   "The Eagle and the Jaguar: Notes for a Bestiary," in *Smoke and Mist: Mesoamerican Studies in Memory of Thelma D. Sullivan* (J.K. Josserand & K. Dakin, eds.), pp. 161–171. BAR International Series 402, Oxford, England.
1988c   "A knife in the Water: The Stingray in Mesoamerican Ritual Life," *Journal of Latin American Lore*, 14:2, pp. 173–191.
In press   "The Lord, The Ruler: Jaguar Symbolism in the Americas," in *Felines: Their Significance, Symbolism and Representation in Precolumbian and Native America*, (Nicholas J. Saunders, ed.). BAR Press, Oxford.
n.d.   "In Love and War: Hummingbird Lore," Latin American Indian Literatures conference (1988), Guatemala City.
n.d.   "The Multimedia Monkey, or, The Failed Man: The Monkey as Artist." Paper delivered at the Seventh Mesa Redonda (1989), Palenque.

Berrin, Kathleen (Ed.)
1978   *Art of the Huichol Indians*. The Fine Arts Museums of San Francisco and Abrams, Inc., New York.

Blaffer, Sarah C.
1972   *The Black-men of Zinacantan*. University of Texas Press, Austin and London.

Broda, Johanna
1987   "Templo Mayor as Ritual Space," in *The Great Temple of Tenochtitlan*, (Broda et al.), pp. 61–123. University of California Press, Berkeley, Los Angeles, London.

Carlson, John B.
1988a   "Skyband Representations in Classic Maya Vase Painting," in *Maya Iconography* (Elizabeth P. Benson and Gillett G. Griffen, eds.), pp. 277–293. Princeton University Press, Princeton, NJ.
1988b   "The Maya Lowlands," in *The Face of Ancient America* (Parsons et al.). Indianapolis Museum of Art and Indiana University Press, Indianapolis.

Cordry, Donald
1981   *Mexican Masks*, University of Texas Press, Austin and London.

Caso, Alfonso and Ignacio Bernal
1952   *Urnas de Oaxaca*. Instituto Nacional de Antropología e Historia, Mexico.

Coe, Michael D.
1972   "Olmec Jaguar and Olmec Kings," in *The Cult of the Feline* (Eliz. P. Benson, ed.), pp. 1–18. Dumbarton Oaks Research Library and Collections, Washington.
1978   *Lords of the Underworld: Masterpieces of Maya Ceramics*. The Art Museum, Princeton University, Princeton, New Jersey.
1988   "Ideology of the Maya Tomb," in *Maya Iconography* (Eliz. P. Benson and Gillett G. Griffen, eds.), pp. 222–235. Princeton University Press, Princeton.

Clancy, Flora S., Clemency Coggins, T.P. Culbert, C. Gallenkamp, P.D. Harrison, & J.A. Sabloff.
1985   *Maya: Treasures of an Ancient Civilization* (C. Gallenkamp & R.E. Johnson, gen. eds.). Harry N. Abrams, Inc., New York.

Durán, Fray Diego
1971   *Book of the Gods and Rites and the Ancient Calendar* (F. Horcasitas and D. Heyden, trans. and eds.). University of Oklahoma Press, Norman.

Dwyer, Jane P. and Edward B.
1975   *Fire, Earth and Water: Sculpture from the Land Collection of Mesoamerican Art*. Fine Arts Museums of San Francisco, San Francisco.

Foster, George M.
1967   "Contemporary Pottery and Basketry," in *Handbook of Middle American Indians*, (Robert Wauchope, gen. ed.), vol. 6, pp. 103–124. University of Texas Press, Austin.

Frost, Gordon
1976   *Guatemalan Mask Imagery*. Southwest Museum, Los Angeles.

Furst, Peter T.
1968   "The Olmec Were — Jaguar Motif in the Light of Ethnographic Reality" in *Dumbarton Oaks Conference on the Olmec*, (Eliz. P. Benson, ed.), pp.143–178. Dumbarton Oaks Reasearch Library and Collections, Washington.
1978   *The Ninth Level: Funerary Art from Ancient Mesoamerica*. Univ. of Iowa Museum of Art, Iowa City.
1981   "Jaguar Baby or Toad Mother: A new look at an old problem in Olmec iconography," in *The Olmec and their Neighbors* (Eliz. P. Benson, ed.), pp.149–162. Dumbarton Oaks Research Library and Collections, Washington.

Gallagher, Jacki
1983   *Companions of the Dead: Ceramic Tomb Sculpture from Ancient West Mexico*. Museum of Cultural History, University of California, Los Angeles.

BULL MASK   96

Maya Indian, Western Highlands, Guatemala
Wood and horn. 10″ × 12″
Collected circa 1970
Gordon Frost Folk Art Collection, Benicia. 134-P4

The mask was probably used in the Dance of the Bulls. The bull was introduced to Mesoamerica by the Spanish and has been incorporated into dance dramas and native religious ritual in various areas of Mesoamerica including Guatemala, Sonora and Chiapas.

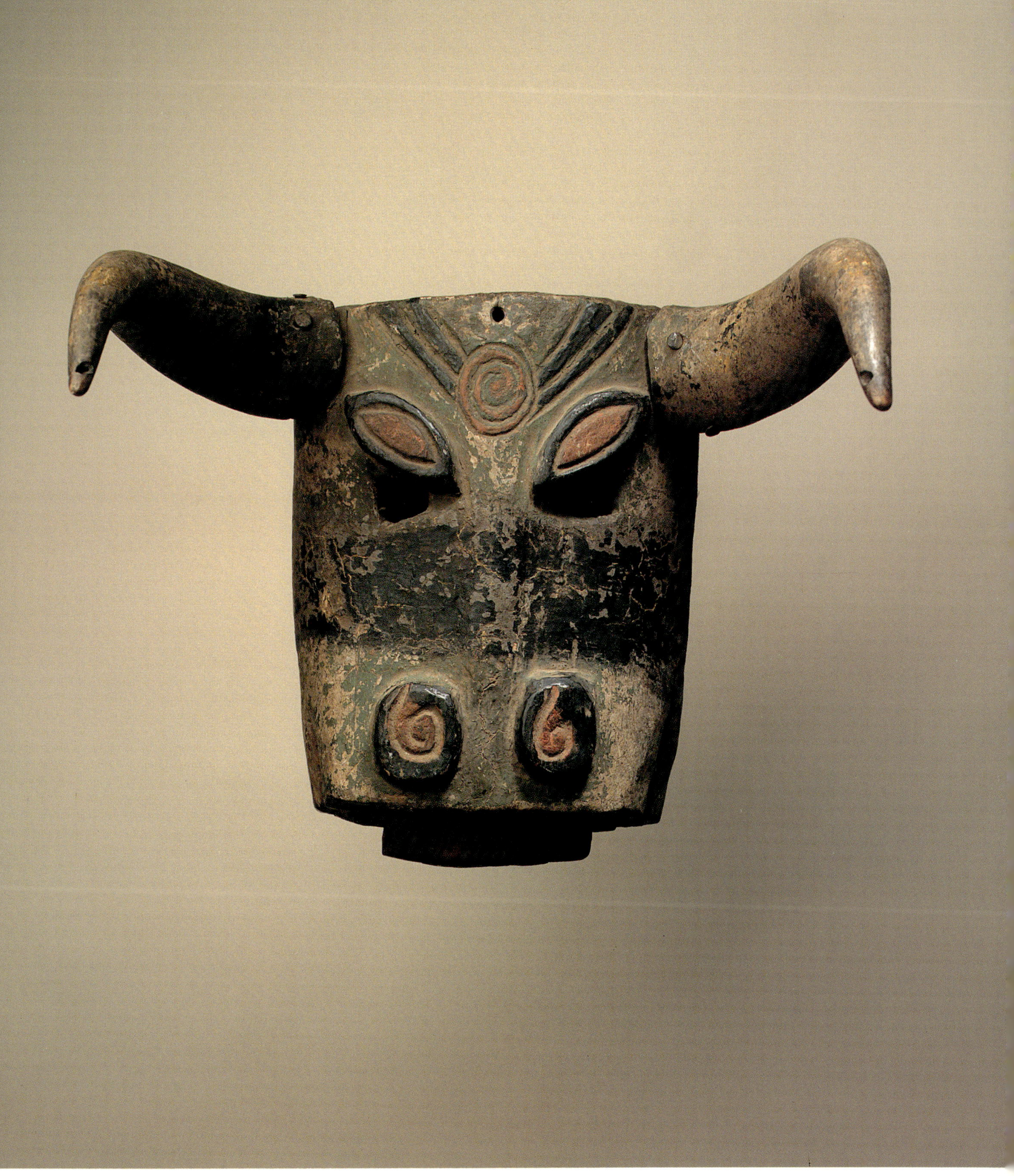

SELECTED BIBLIOGRAPHY, cont.

Gibson, Charles
1964   *The Aztecs Under Spanish Rule.* Stanford University Press, Stanford.

Gossen, Gary
1986   "Mesoamerican Ideas as a Foundation for Regional Synthesis," in *Symbol and Meaning Beyond the Closed Community: Essay in Mesoamerican Ideas,* pp. 1–8. Institute for Mesoamerican Studies: SUNY, Albany.

Green, Judith Strupp
1972   "The Days of the Dead in Oaxaca, Mexico: An Historical Inquiry," *Omega,* vol. 3, n. 3, pp. 245–261.

Hellmuth, Nicholas M.
1987   *Monster and Menschen in der Maya-Kunst.* Akademische Druck-u. Verlagsanstalt, Graz.

Heyden, Doris
1983   *Mitología y simbolismo de la flora en el Mexico prehispanico.* Universidad Nacional Autónoma de Mexico, Mexico.
1986   "Metaphors, Nahualtocaitl, and other 'Disguised' Terms among the Aztecs," in *Symbol and Meaning Beyond the Closed Community: Essays in Mesoamerican Ideas,* (Gary Gossen, ed.), pp. 35–43. Institute for Mesoamerican Studies: SUNY, Albany.

Hunt, Eva
1977   *The Transformation of the Hummingbird.* Cornell University Press, Ithaca and London.

Isbell, Billie Jean
1985   "The Metaphoric Process: From Culture to Nature and Back Again," in *Animal Myths and Metaphors in South America* (Gary Urton, ed.), pp. 285–313. University of Utah, Salt Lake City.

Johnson, Irmgard Weitlaner
1971   "Basketry and Textiles," *Handbook of Middle American Indians,* (Robert Wauchope, gen. ed.), vol. 10, Pt. 1, pp. 297–321. University of Texas Press, Austin.
1976   *Design Motifs on Mexican Indian Textiles.* Vol 1/I–1/II. Akademische Druck-u. Verlagsanstalt, Graz, Austria.

Jones, Tom
1985   "The Xoc, the Sharke, and the Sea Dogs: An Historical Encounter," in *Fifth Palenque Round Table* (Virginia Fields, vol. ed. and M.G. Robertson, gen. ed.), vol. 7, pp. 211–222. Pre-Columbian Art Research Institute, San Francisco.

Joralemon, Peter David
1976   "The Olmec Dragon: A Study in Pre-Columbian Iconography," in *Origins of Religious Art and Iconography in Preclassic Mesoamerica* (H.B. Nicholson, ed.), pp. 29–71. UCLA Latin American Center Publications, Los Angeles.
1988   "The Olmec," in *The Face of Ancient America* (Lee A. Parsons et al.), pp. 9–50. Indianapolis Museum of Art, Indianapolis.

Kan, Michael, Clement Meighan, and H.B. Nicholson
1989   *Sculpture of Ancient West Mexico: Nayarit, Jalisco, Colima: The Proctor Stafford Collection.* Los Angeles County Museum of Art, with Univ. of New Mexico Press, Albuquerque.

Kennedy, Alison Bailey
1982   "Ecce Bufo: The Toad in Nature and in Olmec Iconography," *Current Anthropology* vol. 23, no.3, pp.273–290.

Kerr, Justin
1989   *The Maya Vase Book* (Justin Kerr, ed.), vol. I. Kerr Associates, New York.

Klein, Cecelia F.
1976   *The Face of the Earth.* Garland Publishing, New York.

Labbé, Armand. J.
1982   *Religion, Art, and Iconography: Man and Cosmos in Prehispanic Mesoamerica.* Bowers Museum Foundation, Santa Ana.

León-Portilla, Miguel
1963   *Aztec Thought and Culture,* University of Oklahoma Press, Norman.
1973   *Time and Reality in the Thought of the Maya.* Brown Press, Boston.

Leopold, A. Starker
1959   *Wildlife of Mexico: The Game Birds and Mammals.* University of California Press, Berkeley.

Leyenaar, Ted J.J.
1978   *Ulama: the Perpetuation in Mexico of the Pre-Spanish Ball Game Ullamaliztli.* Leiden, The Netherlands.

Leyenaar, Ted J.J. and Lee A. Parsons
1988   *Ulama: The Ballgame of the Mayas and Aztecs.* Leiden, SMD, The Netherlands.

Lothrop, Samuel Kirkland
1927   "A Note on Indian Ceremonies in Guatemala," *Indian Notes,* vol. 4, no. 1, Museum of the American Indian, New York.
1929   "Further Notes on Indian Ceremonies in Guatemala," *Indian Notes,* vol. 6, no. 1, Museum of the American Indian, New York.

Lumholtz, Carl
1902   *Unknown Mexico.* 2 vols. Charles Scribners and Sons, New York.

Madsen, William
1967   "Religious Syncretism," in *Handbook of Middle American Indians,* (Robert Wauchope, gen. ed.), vol. 6, pp. 369–391. University of Texas Press, Austin.

Markman, Roberta H. and Peter T. Markman
1989   *Masks of the Spirit.* University of California Press, Berkeley and Los Angeles.

Motolinía, Toribio de Benavente
1950   *History of the Indians of New Spain* (Trans. by E.A. Foster). Greenwood Press, Westport.

Nicholson, Henry B.
1971a   "Religion in Pre-Hispanic Central Mexico" in *Handbook of Middle American Indians.* (Robert Wauchope, gen. ed.), vol. 10, pp. 395–446. Middle American Research Institute, Tulane University and University of Texas Press, Austin.
1971b   *Ancient Art of Veracruz.* Ethnic Arts Council of Los Angeles and Los Angeles County Museum of Art, Los Angeles.

Nicholson, H.B. and Alana Cordy-Collins
1979   *Pre-Columbian Art from the Land Collection.* California Academy of Sciences, San Francisco.

SMALL BENCH IN THE FORM OF A DOG   97
Otomí Indian, San Pablito, Puebla
Wood. 24″ × 16″
Collected by Ann Pitzer and Elizabeth Cuellar in the village
Ann Pitzer Collection.

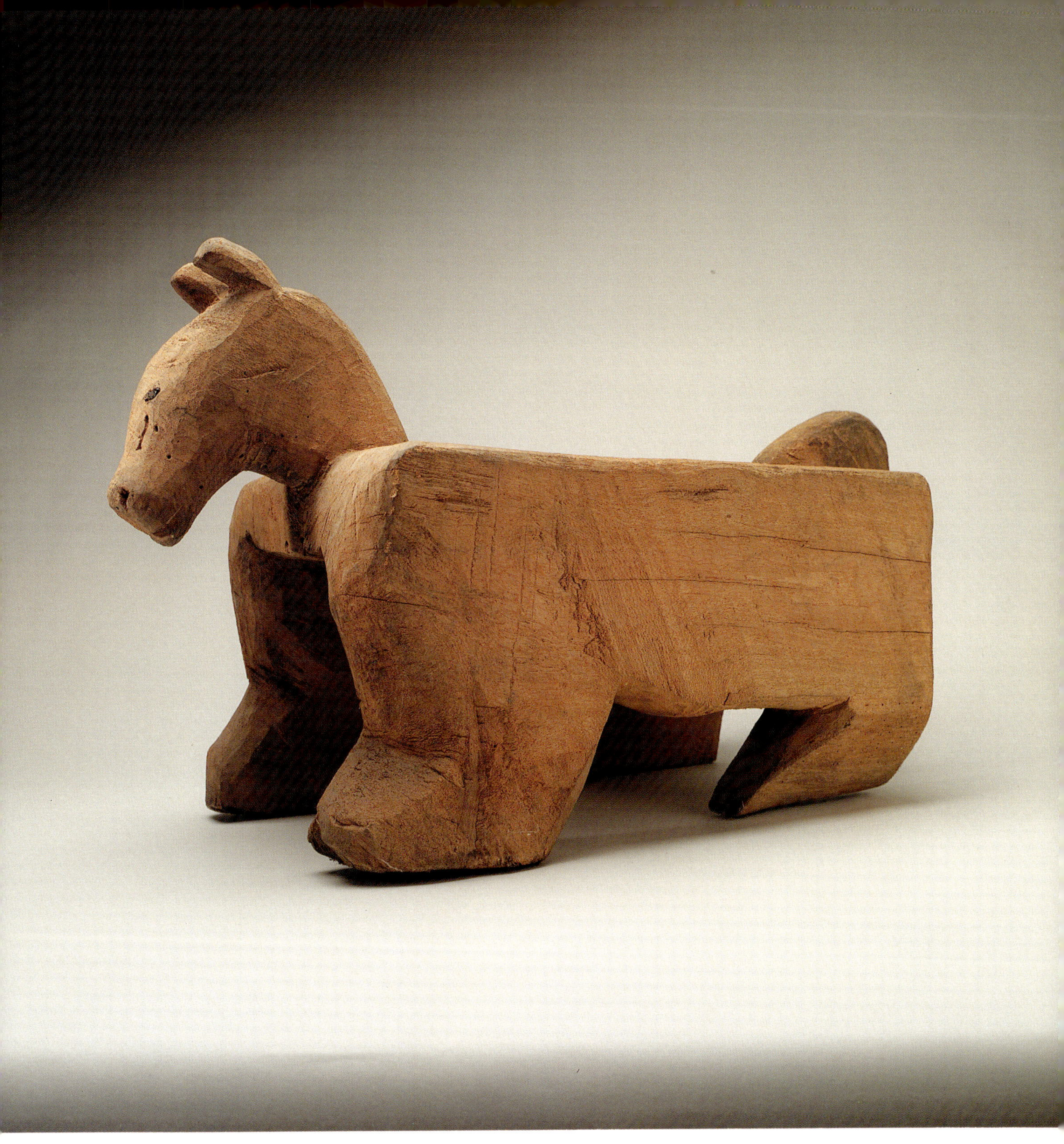

SELECTED BIBLIOGRAPHY, cont.

Nicholson, H.B. and Eloise G. Keber
1983   *Art of Aztec Mexico.* National Gallery of Art, Washington.

Paret-Limardo de Vela, Lise
1963   *La danza del venado en Guatemala,* Centro Editorial "Jose de Pineda Ibarra," Guatemala.

Pasztory, Esther
1983   *Aztec Art.* Harry N. Abrams, Inc., New York.

Peterson, Jeanette F., ed.
1983   *Flora and Fauna Imagery in Precolumbian Cultures: Iconography and Function.* B.A.R. International Series, Oxford.

Puleston, Dennis
1977   "The Art and Archaeology of Hydraulic Agriculture in the Maya Lowlands," in *Social Processes in Maya Prehistory,* (N. Hammond, ed.), pp.449–469. Academic Press, New York.

Real, Katarina
n.d.   "The Armadillo in the Folk Art of the Americas," 1978, ms. in author's possession.

Redfield, Robert
1936   "The Coati and the Ceiba," *Maya Research,* Vol. III, July–Oct.

Redfield, Robert and Alfonso Villa Rojas
1934   *Chan Kom: a Maya Village.* University of Chicago Press, Chicago.

Robicsek, Francis and Donald Hales
1981   *The Maya Book of The Dead: The Ceramic Codex.* University of Virginia Art Museum, Charlottesville.

Sahagún, Bernardino de
1950–1982   *Florentine Codex: General History of the Things of New Spain* (Arthur J.O. Anderson and Charles E. Dibble, trans. and eds.), 13 Vols. School of American Research and University of Utah, Salt Lake City.

Schele, Linda and Mary Ellen Miller
1986   *The Blood of Kings.* Kimbell Art Museum, Fort Worth.

Seler, Eduard
1909   "The Animal Pictures of the Mexican and Maya Manuscripts," in *The Collected Works of Eduard Seler,* vol. 4, pt. 5. Peabody Museum Library, Cambridge.

Shafer, Lynn Cordner
n.d.   "The Function and Meaning of Trees in Postclassic Mesoamerica," 1977, ms. in author's possession.

Starr, Frederick
1899   *Indians of Southern Mexico: An Ethnographic Album,* Published by the author, Chicago.
1902   *Notes Upon the Ethnography of Southern Mexico,* vol. 8, part 1, and vol. 9, part 2, Davenport, Iowa.

Stuart, David
1989   "Hieroglyphs on Maya Vessels," in *The Maya Vase Book* (Justin Kerr,ed.), vol. I, pp.149–160. Kerr Associates, New York.

Tedlock, Barbara
1982   *Time and the Highland Maya.* University of New Mexico Press, Albuquerque.
1983   "A Phenomenological Approach to Religious Change in Highland Guatemala," in *Heritage of Conquest* (Carl Kendall et al., eds.), pp. 235–246. University of New Mexico Press, Albuquerque.

Tedlock, Dennis
1985   *Popol Vuh.* Simon and Schuster, New York.

Thompson, John Eric Sidney
1970   *Maya History and Religion.* University of Oklahoma Press, Norman.

Toor, Frances
1947   *A Treasury of Mexican Folkways.* Crown Publishers, New York.

Tozzer, Alfred M., Ed.
1941   *Landa's Relación de las cosas de Yucatan.* Papers of the Peabody Museum of American Archaeology and Ethnology, vol. 18. Harvard University Press, Cambridge.

Urton, Gary, Ed.
1985   *Animal Myths and Metaphors in South America.* University of Utah Press, Salt Lake City.

Villa Rojas, Alfonso
1969   "The Maya of Yucatan," in *Handbook of Middle American Indians* (ed. Robert Wauchope), Vol. 7, pp. 244–275. University of Texas Press, Austin.

Vogt, Evon Z.
1969   *Zinacantan: A Maya Community in the Highland of Chiapas.* The Belknap Press at Harvard University Press, Cambridge.
1976   *Tortillas for the Gods: A Symbolic Analysis of Zinacanteco Rituals.* Harvard University Press, Cambridge.

Von Winning, Hasso, and Alfred Stendahl
1968   *Pre-Columbian Art of Mexico and Central America.* Harry Abrams, Inc., New York.

WOMAN WITH BIRDS AND FLOWERS   98
Oaxaca, Atzompa, Mexico
Terracotta sculpture by Teodora Blanco
Collection of Mingei International
Gift of Fred and Barbara Meiers

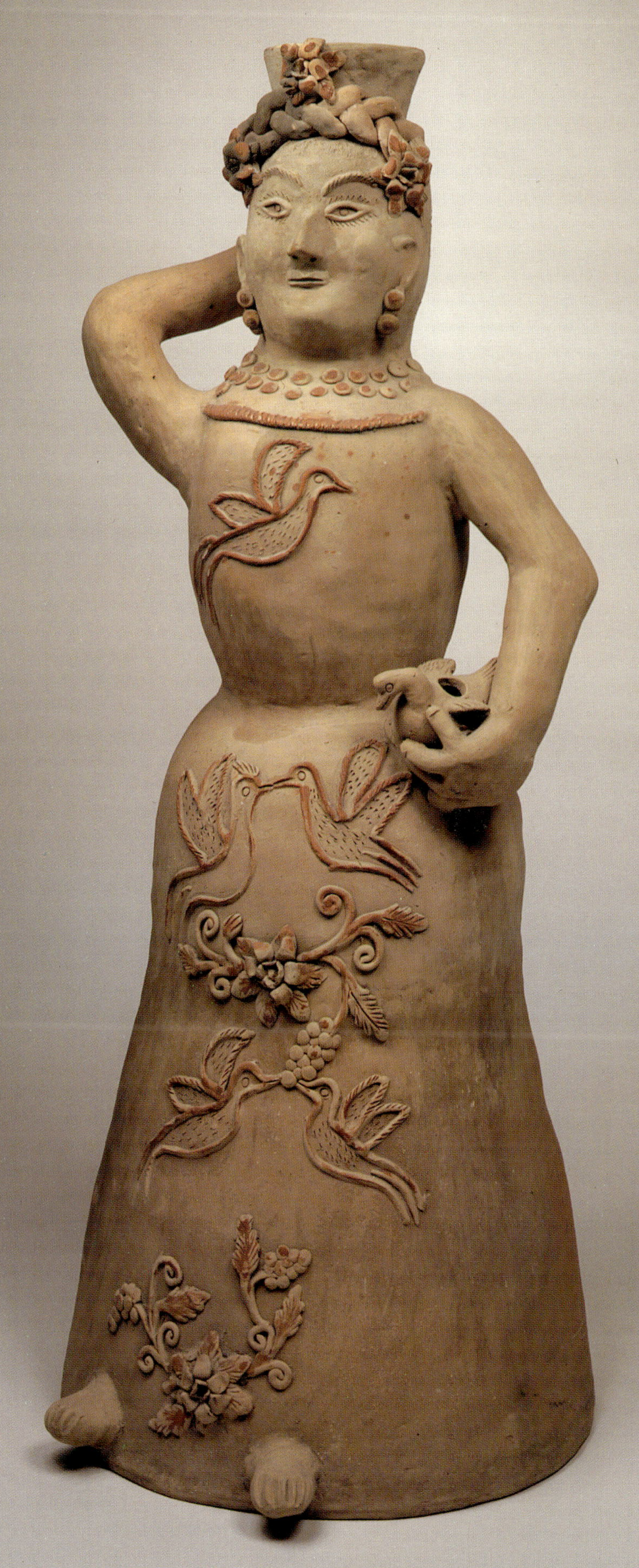

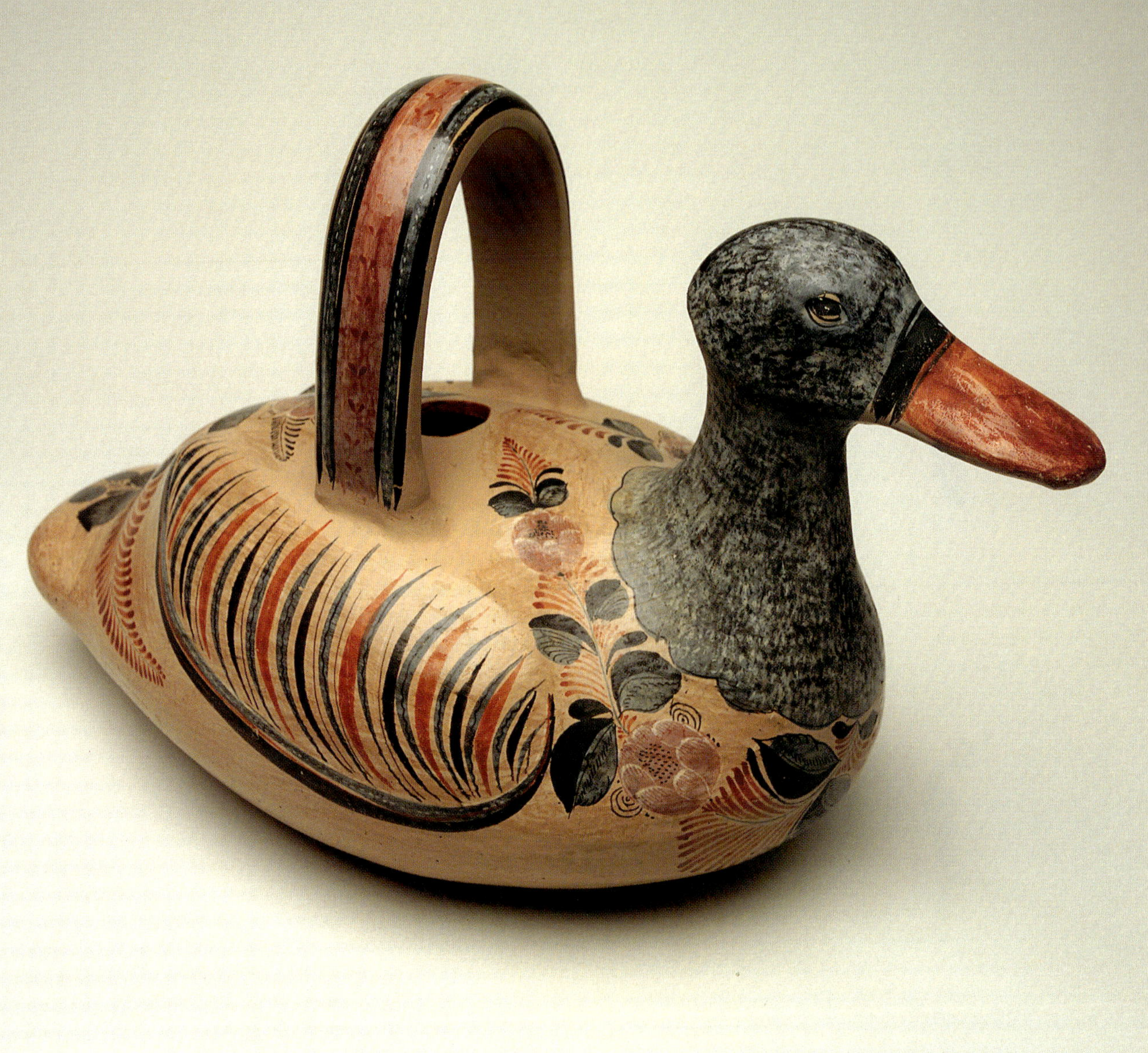

VESSEL IN THE FORM OF A DUCK   100
Mexican, Tonalá, Jalisco, Mexico
Burnished earthenware of the type known as
"de olor" (scented). 14½" × 8¼"
The Lowie Museum of Anthropology, University
of California at Berkeley 3-15753

Duck shaped vessels (*cantimploras*) were used to
hold holy water on home altars since at least the last
century.

VESSEL IN THE SHAPE OF A   99
GOURD WATER CONTAINER
Mexican
Tonalá, Jalisco, Mexico
Burnished and painted earthenware with a corn cob
stopper. 9" × 13"
Collected in 1967
San Diego Museum of Man 1967-71-49

Terracotta vessels made in the form of gourds were
common before the Conquest. This example of burnished ware was made by Tonala artisan Amado Galvan.

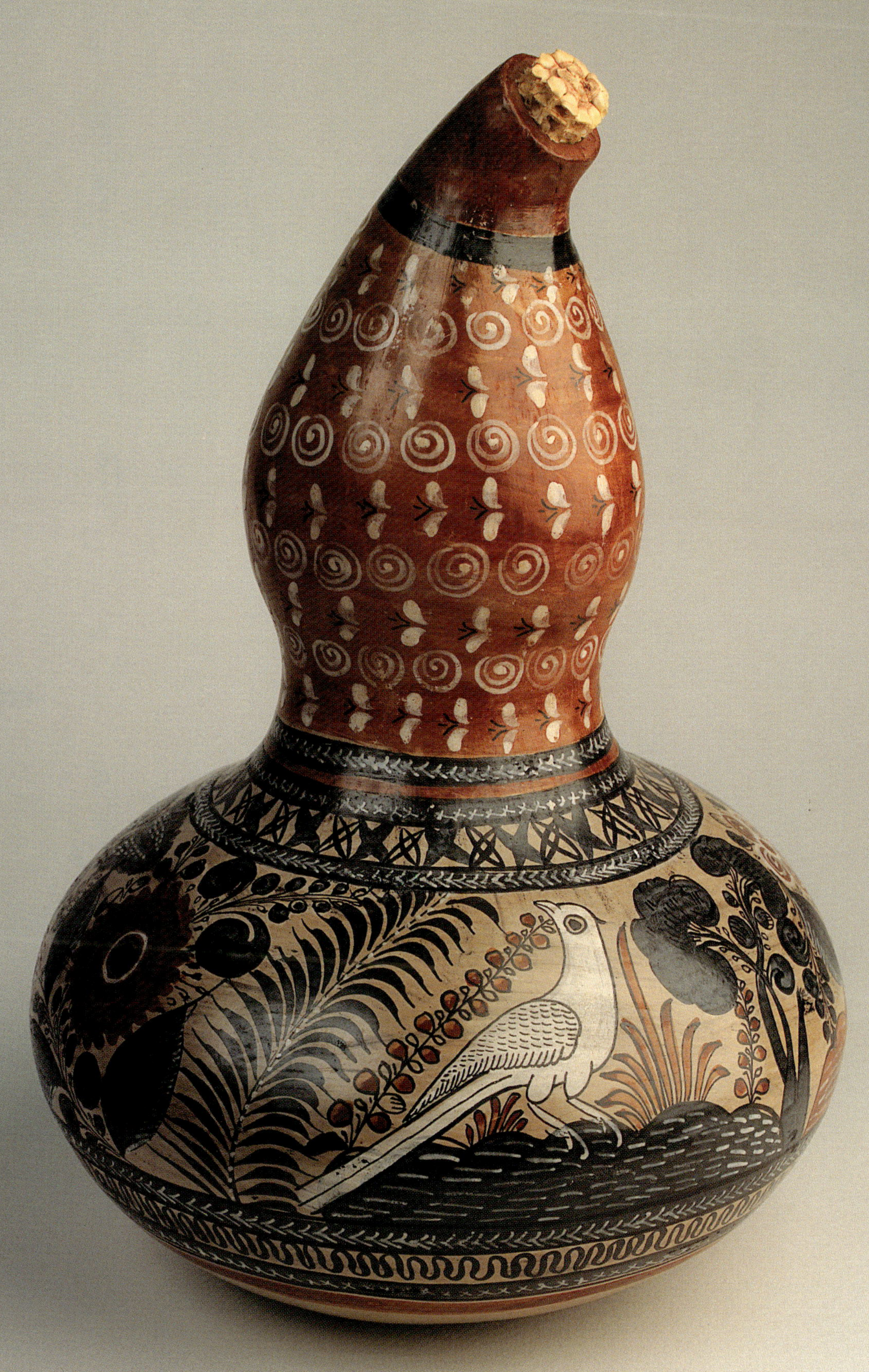

MINGEI INTERNATIONAL MUSEUM

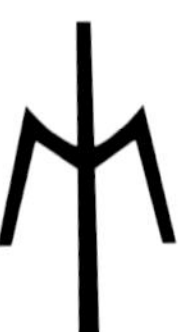

Board of Directors

JEAN HAHN, President
ALAN JAFFE, Vice President
KATY DESSENT, Secretary-Treasurer
DOROTHY D. STEWART, Corresponding Secretary
ROGER C. CORNELL, M.D.
SHIRLEY KOVAR
MARTHA LONGENECKER
ALTHEA D. LUCIC
DIANE POWERS
DAVID RAPHAEL SINGER
BRADLEY SMITH
HAROLD K. TICHO

Development Consultants of the Board

SALLY DRACHMAN, Ph.D.
WILLIAM W. GORDON
ROBERT E. CORDER

Emeritus Directors

ALEX DE BAKCSY
LESLIE HINTON
SAM HINTON
BARBARA MEIERS
FRED MEIERS
JAMES F. MULVANEY
DAVID RINEHART

JAMES DAWE, Chairman of Development Board 1980–1984
SYDNEY MARTIN ROTH, Management Consultant 1977–1986

International Advisory Board

REAR ADMIRAL AND MRS. W. HALEY ROGERS, Honorary Chairmen

DR. BERNARD W. AGINSKY, anthropologist and Founder of The Institute
for World Understanding — Peoples, Cultures, and Languages
LAURA ANDRESON, emeritus professor of art, UCLA
JAN ARNOW, Writer, Southern American crafts
ELEANOR BEACH, American art lecturer & photographer
WARREN BEACH, former director of the Fine Arts Gallery of San Diego
ROBERT BISHOP, Director of the Museum of American Folk Art, New York
MRS. FRANK BUTLER, Lawyer
RAND CASTILE, director of The Asian Art Museum, San Francisco
KATARINA REAL-CATE, anthropologist, South American folk art
MRS. JOHN COLE, bibliographer of folk art
NORMAN COUSINS
ELIZABETH CUELLAR, curator, International Folk Art Collection,
University of Mexico
THE HONORABLE AND MRS. WARREN M. DORN

SAM AND LESLIE HINTON, Folklorists
ROBERT BRUCE INVERARITY, museum director & consultant
DR. STEFAN MAX, European linguist and cultural liaison
FRED AND BARBARA MEIERS, Mexican Folk Art scholars
M.C. MADHAVAN, Ph.D., International Development
JOHN DARCY NOBLE, Curator Emeritus, Museum of the City of New York
JOHN NORTON, honorary consul of Sweden
JAMES S. PLAUT, art historian, cultural administrator
TSUNE SESOKO, director, Cosmo Public Relations
HAKU SHAH, curator of the Tribal Research Museum, India
TATSUZO SHIMAOKA, one of Japan's foremost potters
BEATA SYDOFF, Cultural Councellor of Sweden
LENNOX TIERNEY, professor, curator of Oriental art, University of Utah
OPPI UNTRACHT, art writer, Finland
CATHERINE YU-YU CHO WOO, Chinese scholar and artist
SORI YANAGI, designer, son of founder of Mingei Association of Japan

Legal Counsel

SHIRLEY KOVAR, Gary, Cary, Ames & Frye

C.P.A.

BRUCE HEAP, Hutchinson & Bloodgood, C.P.A.

Staff

MARTHA LONGENECKER, Director
JAN JENNINGS, Director's Assistant
JULIA BRASHARES, Registrar
CORNELIA FEYE, Assistant Registrar
NAN DANNINGER, Publications Marketing
DOROTHEA CRONOGUE, Publicist
ERIC JARVIS, Membership Coordinator
ROSEMARY WILSON, Operations Assistant
JOYCE CORBETT, Collectors' Gallery Coordinator
ANNA SAULSBERY, Collectors' Gallery Bookkeeper
ARLENE EDING, Docent Coordinator
SUSAN EYER, Mailing Coordinator
BETTY GRENSTED, Librarian
PATTY GORMAN, Excursions Coordinator
EILEEN MILLER, Receptions Coordinator
VIRGINIA PANTONE, Volunteer Coordinator
ANNE WEAVER, Collections Conservator

Principal Museum Founding Benefactor

ERNEST W. HAHN

Major Museum Founding Benefactors

THE JAMES IRVING FOUNDATION
THE PARKER FOUNDATION

Distinguished Museum Founding Benefactors

MRS. JARVIS BARLOW
THE KNIGHT AID FUND

Museum Founding Benefactors

CALIFORNIA FIRST BANK
LAS PATRONAS

Museum Founding Life Members

ERNEST W. HAHN
LOUISA S. KASSLER
DRS. ROBERT and MARY KNIGHT
MRS. L.E. PETERSON
ROBERT O. PETERSON
SYDNEY MARTIN ROTH
MR. and MRS. HORTON R. TELFORD
MR. MASAO TSUYAMA
PHYLLIS K. WALKER and DAVID C. WALKER

Museum Life Members

MARC APPLETON
MRS. JARVIS BARLOW
ROSEMARIE BRAUN
BRUCE HENDERSON
MR. and MRS. R.C. MATHES
DR. LOUISE HAWKES PADELFORD
DAVID RINEHART
REAR ADMIRAL and MRS. FORREST ROYAL
SEATTLE OFFICE SERVICE, INC.
MR. and MRS. WORLEY W. STEWART

Patron Members

MR. and MRS. FRANK BUTLER
DR. and MRS. KIRK J. DAVID
MR. and MRS. ROSCOE E. HAZARD, JR.
MR. KENNETH E. and MRS. DOROTHY V. HILL
MS. REBECCA WOOD
MR. BERNARD SMOLIN

Director's Circle

ERNEST W. and JEAN HAHN, Honorary Chairmen
DR. BERNARD W. AGINSKY
FRANCES ARMSTRONG
BARBARA BAEHR
CAROLE A. BRANSON
BETTY BUFFUM
MR. and MRS. J. DALLAS CLARK
DAVID COPLEY
ROGER C. CORNELL, M.D.
SUSAN CRUTCHFIELD
MR. and MRS. MICHAEL DESSENT
MR. and MRS. EDWARD DRCAR
DR. and MRS. CHARLES EDWARDS
DR. and MRS. LAURENCE FAVROT
MR. and MRS. JOSEPH L. FRITZENKOTTER
MR. and MRS. JOHN E. GOODE
DR. and MRS. HUBERT GREENWAY
MR. and MRS. BRUCE HEAP
JOAN HOLTER
ALAN and NORA JAFFE
LOUISA KASSLER
MR. and MRS. JAMES KERR
DRS. MARY and ROBERT KNIGHT
SHIRLEY KOVAR
MR. AND MRS. FRED MARSTON
DR. and MRS. KIRK PETERSON
CAROLYN PITCAIRN
ROBERT and DIANE POWERS
DR. and MRS. ROGER REVELLE
MR. and MRS. GEORGE P. RODES
REAR ADMIRAL and MRS. W. HALEY ROGERS
SYDNEY MARTIN ROTH
ANNA M. SAULSBERY
MRS. THOMAS SHEPHERD
DR. BERNARD SIEGAN
BRADLEY SMITH
MR. and MRS. WORLEY W. STEWART
MR. JULES S. TRAUB
BARBARA WALBRIDGE
MR. and MRS. FRANK R. WARREN
LORI WINTER
MR. and MRS. WALTER ZABLE

Corporate Associates

Catellus Development Corporation
The James S. Copley Foundation
JMB Realty Corporation

CREDITS

Exhibition

JEANETTE F. PETERSON, guest curator
JUDITH STRUPP GREEN, curatorial consultant
MARTHA LONGENECKER, design
JO ANNE HEANEY, co-designer
JULIA BRASHARES, registrar
CORNELIA FEYE, assistant registrar
TRACY BROWN, graphics

Installation
ERIC JARVIS
ROSEMARY WILSON
CARMELLA CALDERA
MARTHA EHRINGER
MIRIAM ELLIOT
MYRON GORDON
SCOTT JAFFE
TONY RAZCKA
NORMA AVENDANO
KIM DUCLO
LUTES CABINETS
and the VOLUNTEER COUNCIL

Publication

MARTHA LONGENECKER, design
JAN JENNINGS, editing
MOOG AND ASSOCIATES INC., typesetting
  Dee Kahler, typographer
KATHY JOHNSON, production
SPRING COLOR, color separations
  Rep. Tim Gallagher
VANARD LITHOGRAPHY, printing
VICTOR ARREOLA, production assistant

Photography
LYNTON GARDINER

Additional Photographs Courtesy of:
DENVER ART MUSEUM
  figures: 5, 18, 19, 30, 32, 42, 55, 63, 68, 69, 72
FOWLER MUSEUM OF CULTURAL HISTORY, UCLA
  page 37
LOS ANGELES COUNTY MUSEUM OF ART
  figures: cover, frontispiece, title page, 21, 28, 38, 48, 62
ROLANDO ROSITO
  figures: 1, 25, 35, 49, 54, 65, 76
GORDON FROST
NICHOLAS HELMUTH
GRACE JOHNSON
JUSTIN KERR
NATIONAL GEOGRAPHIC SOCIETY
SAN DIEGO ZOOLOGICAL SOCIETY

ORCHESTRA
by Teodora Blanco, 1965
Oaxaca, Atzompa, Mexico
Clay musicians. H: 2–4″
Collection of Mingei International

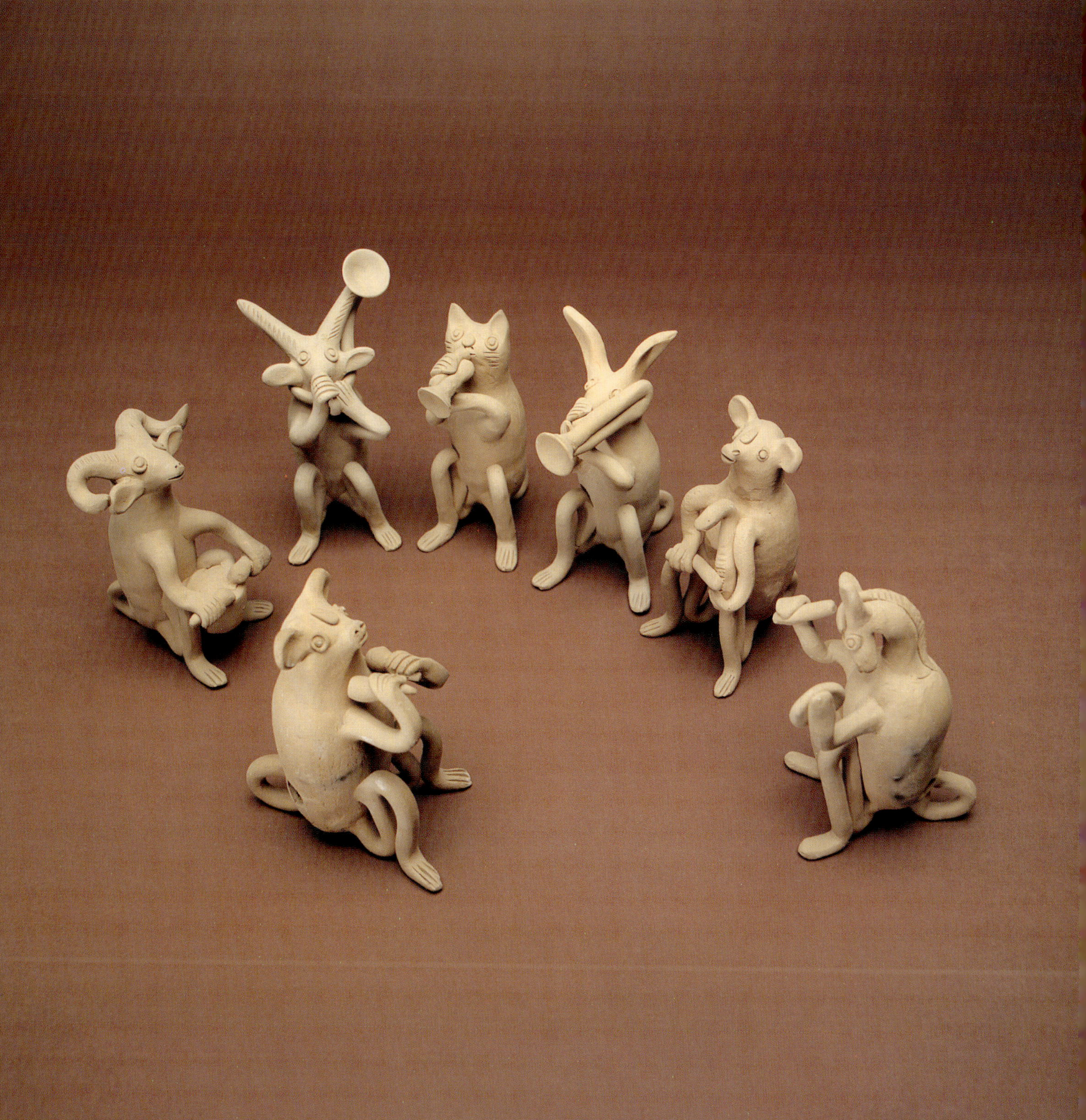

MINGEI INTERNATIONAL was incorporated in 1974 as a non-profit public foundation dedicated to furthering the understanding of world folk art.

These are essential arts of people living in all times throughout the world that share a direct simplicity and reflect a joy in making, by hand, useful objects that are satisfying to the human spirit.

MINGEI INTERNATIONAL built and established the MUSEUM OF WORLD FOLK ART at University Towne Centre, La Jolla, California, on May 5, 1978. Here the art of the people, by the people and for the people comes to the people.

Through the universal language of line, form and color the arts of the people speak for themselves and inspire appreciation of the similarities and distinctions of individuals and cultures.

MINGEI is a special word used transculturally for ''art of the people.'' It was coined in the early twentieth century by the revered scholar, Dr. Soetsu Yanagi, who combined the Japanese words for *all people,* MIN, and *art,* GEI. His keen eye observed that many articles made by unknown craftsmen of pre-industrial times were of a beauty seldom equaled by artists of modern societies.

From questioning why this might be, he gained insight as to the nature of beauty embodied in objects which are integrally related to life and born of a state of mind not attached to a conscious idea of beauty or ugliness.

Within these timeless arts of the people (Mingei), he recognized a quality of expression in which there was no fragmentation of body, mind and spirit. He realized that to balance the weight of increasing technology there is a growing urgency for man to continue to make and use objects that express his whole being.

To communicate this profound insight, Dr. Yanagi and the renowned potters, Shoji Hamada and Kanjiro Kawai, founded the Mingei Association and first Folk Art Museum of Japan.

Inspired by — but not affiliated with that organization — Mingei International Museum opens a window to a broad and intimate view of our magnificent world through the timeless arts of the people.

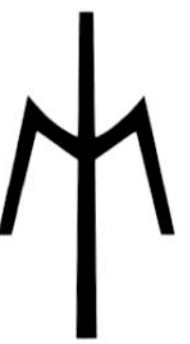